Project Communication Management in Complex Environments

Zhong Ying · Low Sui Pheng

Project Communication Management in Complex Environments

Springer

Zhong Ying
Low Sui Pheng
Department of Building
National University of Singapore
Singapore
Singapore

ISBN 978-981-10-1210-5 ISBN 978-981-4560-64-1 (eBook)
DOI 10.1007/978-981-4560-64-1
Springer Singapore Heidelberg New York Dordrecht London

Printed on acid-free paper

Springer is part of Springer Science+Business Media (www.springer.com)

Preface

The number of crisis incidents and their severity is rising along with the growing complexity of technology and society. In today's dynamic, high-velocity social, and business environment, which is characterized by discontinuity and continuous change, crises are understood as more the norm rather than the exception in organizations. Managers increasingly realize that at any point in time even when we are not in a crisis, we are nevertheless in a pre-crisis mode. Construction is typically a complex, crisis-prone activity carried out in an environment that is relatively uncontrollable, compared to the manufacturing industry. There are many incidents and crises that can interrupt the progress in construction projects. Crises described as high-impact crisis events create high levels of threat and uncertainty to crisis management and communication in construction.

The crisis response phase puts the organization's established normal communication systems and processes under enormous and additional pressure. Conventional crisis response communication models and management grounded on the command-and-control principles of "scientific management" focus on careful planning and break the crisis down into discrete stages that follow a linear sequence, with a tendency to move toward a state of stable equilibrium. Although there may be an effective tool for structuring problems within a known set of options, this is rather limited in describing flexible reactions to the rapidly changing circumstances and explaining the dynamic and complex crisis response situations.

Complexity theory developed from systems theory and, much like systems theory, can be applied to a wide spectrum of disciplines. It has enriched many areas of inquiry by expanding the applications of the systems perspectives to the nonlinear operation of complex systems, and making a paradigm shift in postulating that the forces of disorder, instability, nonlinearity, and unpredictability are controlling the universe. Through this complex view, crisis response communication system can be viewed as a complex adaptive system. The behavior of these complex systems cannot be predicted; all the parts self-organize and learn and adapt to their dynamically changing environment. Hence this book proposes that the purpose of the crisis response communication system needs to be redefined by complementing it with a broader concept of complexity theory to enhance the organization's adaptability and resilience in the event of crisis.

The objectives of this book are to make a contribution to link and extend the knowledge of complexity theory on communication management in the context of the crisis response. Drawing upon relevant concepts in organizational studies and management and complexity theory, this book proposes and refines a conceptual framework for understanding the underlying pattern of communication behavior and decisions of human systems in response to a crisis, and develops a complexity-informed model for effective crisis response communication management for construction organizations. To achieve these objectives, questionnaire surveys and interviews with 46 Chinese construction firms faced with a major natural crisis during the period of the massive earthquake on May 12, 2008 in urban Chengdu, China were carried out to identify and evaluate the significant organizational variables which influence the control level of the communication system for the construction firms in order to be able to creatively respond to unforeseen events and overcome adaptive challenges. In addition to the data collected from the survey, for validation, two case studies were investigated and analyzed through a formal set of in-depth interviews with company management representatives directly involved in the crisis response and through analysis of secondary data. This book therefore reflects real-life practices as experienced by organizations and people faced with a crisis in the construction industry.

The set of findings presented in this book suggest that the complexity-informed framework can provide a good analytical foundation of providing a new mental model for a deeper understanding of organizational communication behavior and decision-making practice in response to a crisis situation. A critical role for adaptive and flexible crisis response communication management is to make sense of the complexity of the organization's internal and external environment and obtain insights into the significant organizational variables that affect performance. By doing so, an organization has the potential to be able to self-organize and enhance its adaptability and resilience in the event of a crisis and contribute to long-term business continuity management.

In essence, this book highlights the need for successful project communication management in complex environments brought about by crisis (including natural disasters) and how organizations can respond appropriately to such crisis based on pro-active preparations anchored on the complexity-informed framework.

Contents

Chapter 1
Introduction

1.1 Background

In today's dynamic, high-velocity social and business environment, which is characterized by discontinuity and continuous change, crises are understood as more the norm rather than exception in organizations (Paraskevas 2006). Managers increasingly realize that "anytime you are not in a crisis, you are instead in a pre-crisis, or a prodromal mode" (Fink 1986, p. 5). From the perspective of management research, crisis is defined as a low-probability, high-impact event that threatens the viability of organizations and is characterized by ambiguity of cause, effect, and means of resolution, as well as by a belief that decisions must be made swiftly (Pearson and Clair 1998). While the crisis can represent a serious threat to organizational high priority goals, it is concerned with psychological issues that beset managers facing a crisis: extreme time pressures to act, lack of clarity about what is the best action to take, and an element of surprise (Lerbinger 1997).

In recent years, researchers and practitioners have explored the nature, theory, and best practices that are required for effective crisis preparation and response (Coombs 2007). The consequences of being unprepared to respond quickly, appropriately, and ethically to a crisis are dramatic and well documented (Millar and Heath 2003). Crisis can harm the organization's efforts to create and maintain mutually beneficial relationships with interested parties, damage organization's reputation, weaken organization's capability for business continuity management, and result in actual or negative consequences for the health, safety, and well-being of organizations, members and the environment.

Crisis can happen to any organization, and crisis management is therefore crucial for all organizations (Low et al. 1999). When an organization is threatened by external environment crises (e.g., natural disasters) or internal events (e.g., structural changes), the need for communication increases to some extent. Managing these situations requires managers to be highly skilled communicators and negotiators capable of managing various stakeholder expectations and creating a positive and learning culture throughout the organization. Crane and Livesey

Z. Ying and L. Sui Pheng, *Project Communication Management in Complex Environments*,
DOI: 10.1007/978-981-4560-64-1_1,

(2003) highlighted the central role of communication in constituting, managing, and maintaining various stakeholder relationships. Communication is often considered as a background or underpinning context in determining organizational performance, conceptually related to the structure, culture, leadership, and rewards of an organization (Church 1996). "Excellent" organizations use the potential of communication management to assist in transformation and relationships with the environment (Grunig and White 1992).

In the construction sector, the construction project environment presents a particularly problematic arena within which to apply communication practices proven to be effective in other sectors (Dainty et al. 2006). The construction organizations used to procure buildings are project-based and consist of people and stakeholders with diverse and often competing needs who govern the success of construction process. For many years, poor communication practices have been recognized as a serious delimiting factor within the construction industry (Boyd and Wild 2003). Effective organizational communication is challenging as there exist structural constrains and cultural barriers to manage and transfer information across project and professional boundaries (Emmitt and Gorse 2003).

Furthermore, the nature of construction activities creates an uncertainty and hostile environment for construction projects and organizations. Crises seem inevitable in projects, and come from a variety of sources that can interrupt processes in construction projects. Consequently, construction companies that deal with projects on an ongoing basis need to learn to deal with crises on a regular basis (Hallgren and Wilson 2008).

It is suggested that effective communication management is a critical tool in the management of a crisis situation. Grunig and White (1992) proposed a two-way, symmetrical communication model for this purpose. Seeger et al. (2001) emphasized that good communication management following a crisis plays an integral role in the success of crisis management. Ulmer (2001) argued that organizations establish value positions on issues of importance and work to establish instrumental communication channels with various stakeholders in the pre, mid, and post crisis. There has also been considerable evidence in the social and behavior sciences to indicate that the patterns of communications and behavior which emerge in response to a crisis would influence the efficiency of crisis management (Leavitt 1951; Shaw 1981; Loosemore 2000; Carroll and Burton 2000). The benefits of developing and managing effective communication have profound implications for crisis-struck organizations and it can play an important role in how the construction organizations resolve crises they cannot avoid.

1.2 Research Problem

The number of crisis incidents and their severity is rising along with the growing complexity of technology and society (Lerbinger 1997). However, establishing a favorable communication system that is receptive to critical messages and that recognizes complexity and uncertainty is difficult to achieve (Weaver 2007). There have been significant barriers to communication with construction project organizations and informal communication channels are likely to be widely used during a crisis (Bennett 1991). The problems of communication also emerged because of the difficulties in handling large amounts of information generated during a crisis. Crises create ambiguity and uncertainties and a critical need for almost immediate, accurate information provided by the relevant parties. Paradoxically in attempting to cope, people always had a tendency to exhibit extremes of behavior, accumulating pressure, stress, and poor communication which made the situation even worse (Loosemore 1998a).

Typical crisis definitions feature the dynamics and complexity of a crisis, but rarely go into the details of communication options and functions that are required during different stages of a crisis. The approaches and methods entailed in genuine communication still remains not well understood, neither are the implications for organizational actions (Foster and Jonker 2005). Communication writers and theorists have traditionally drawn upon the analogy between the human communication process and the electronic telecommunications process where information in the form of acoustic or visual messages is transmitted or conveyed between parties (Shannon and Weaver 1949; Emmitt and Gorse 2003; Dainty et al. 2006). The most common communication model is developed in which the messages are sent from the "senders" to the "receivers" or "audiences," with "feedback" message from the "receivers" to the "senders" to improve and adjust their messages (Crane and Liversey 2003). It is designed to ensure the situation that one communication party can control the messages and persuade the other party, and the other party can receive the message accurately or as intended.

These simplistic and linear models of communication process have traditionally been applied in the unique sociocultural context of construction to allow the analysis of the processes and semiotics involved (Emmitt and Gorse 2003). Within the context of a crisis situation, conventional crisis communication plan or crisis management plan (CMP) includes "a series of checklists or a template" (Thayer 1998, p. 12) which can help move the construction organization into auto-pilot; communicate proper information to respond to both internal and external stakeholders, including employees, contractors/sub-contractors, financers, government authorities, media, and the press. This approach to crisis management focuses on careful planning and breaks the crisis down into discrete stages that follow a linear sequence (Coombs 2007; Fearn-Banks 2007). Conventionally important advice to organizations has been to plan and prepare actions carefully, and to foresee the consequences of a crisis. The aim has thus been to reduce the complexity and uncertainty in a situation, by applying an "information engineering approach" to crisis planning that sees "crises as objective events whose meaning is both predetermined and self-evident" (Hearit and Courtright 2004, p. 204).

However, more recent research has shown that if too much energy is put into preparation and planning, there are great risks of organizations getting into a deadlock situation. This can cause organizations to become locked into a number of fixed ways of responding and behaving. Consequently, organizations turn to formalizing and centralizing decision making during the crisis. Overly rigid crisis planning procedures can also raise false expectations among managers that make communication less effective (Gilpin and Murphy 2008). Moreover, another problem with traditional crisis planning is that the crisis management team members or personnel rarely possess all the necessary information. Human decision makers have limited cognitive capacity due to the unavoidable imperfection of man's knowledge (Hayek 1954). For example, many organizations make positive assumptions about their key stakeholders only to find they were mistaken when hit by a crisis. In addition, crisis goals are often multiple: resolving a crisis may mean having to satisfy various stakeholders, take corrective actions, limit negative public image, and so on, a complex array of objectives that makes evaluation difficult.

In general, these communication models are derived from the "command and control" principles of "scientific management" (Taylor 1967), with the tendency to move toward a state of stable equilibrium (Stacey et al. 2000). Within this schema, managers tend to control the dynamics within the system, based on linear causality, assuming the roles of task definition and boundary control, and taking timely action to correct for change to preserve equilibrium. The feedback within the system does cause changes in pattern of behavior within the organization, but it does not change the dynamics themselves; the system continues to operate in essentially the same manner as before, simply with more information (Stacey et al. 2000). Although these linear-based communication models may be effective tool for structuring problems within a known set of options, they may be inadequate and rather limited in describing flexible reactions to the changing circumstances and explaining the dynamic and complex crisis response situation.

The purpose of the crisis response system needs to be redefined by replacing the narrow objectives of the crisis management planning's tasks with a broader concept of the organization's adaptability and resilience in the complex environment. The response system should enable the project organization to become resistant to perturbations and enhance its capacity to restore itself after a crisis (Paraskevas 2006). Hence, the management of the complex communication system in response to a crisis requires a new framework that complements the idea of a predictable world operating as a linear cause-and-effect system.

Complexity theory that has grown enormously since the mid-twentieth century has emerged to provide a new, holistic approach that helps to understand the social behaviors of the human system (e.g., project team, organization) and the connection and interaction networks between project team members, organizational members, and project stakeholders under complex and changing circumstances (Anderson et al. 1999; Boisot and Child 1999; McKelvey 2003; McMillan 2006; Stacey 2001). Over the past years, there has been an increasing tendency to draw attention to the particular challenges posed by complex projects (Williams 1999; Richardson et al. 2005) or by complexity in projects (Baccarini 1996; Cicmil 2003; Sommer and Loch 2004). With its focus on uncertainty and unpredictability and organizational relationships, complexity theory offers particular relevance to the crisis management and communication process (Gilpin and Murphy 2008).

It thereby addresses how individuals and organizations are able to acquire and transfer knowledge and information, make sense of and make decisions for the unexpected and rapidly changing crisis environment.

Lewin (1993, p. S10) quoted from Stuart Kauffman that "If complexity theory is valid, ecosystems, economies, or nation states each interact with their world by a similar underlying mechanism—adaptation to the edge of chaos." Theories, implications, and applications of complexity and complex adaptive systems have attracted widespread attention among organizational theorists in the management research arena. One of the most well-known complexity science research community is the Santa Fe Institute (SFI) (available at http://www.santafe.edu/) which was founded in 1984 in New Mexico, with the mission to foster a multidisciplinary collaboration in the physical, biological, computational, and social science and to uncover the mechanisms that underlie the deep simplicity present in our complex world. In Singapore, a proposal for a new Asian version of the SFI for promoting complexity study was made recently in February 2009 (The Straits Time 2009, please see Appendix A).

Another famous research group led by Professor Ralph D. Stacey from the Complexity and Management Center at the University of Hertfordshire Business School in the UK made extensive studies and publications on organizational practice and business management using a complexity theory perspective. For the construction industry, a task group TG62 of International Council for Research and Innovation in Building and Construction (CIB) was established in 2006 with the emphasis on the necessity for a strong, coherent, and international research strategy to address and embrace complexity science within the built environment disciplines.

In an attempt to achieve a more comprehensive appreciation of communication practice in the situation of a crisis, this research examines the implications of complexity theory with linkage to social systems and organization studies to crisis response communication management. It is believed that the concepts of complexity science would provide another view of the communications and interactions between various parts in the crisis response system. By exploring human communication and interaction processes within the construction organizational context, this would allow a more holistic understanding of how communication can be managed more effectively and adaptively in response to a crisis.

1.3 Knowledge Gap

Scholars have long recognized the important role communication plays in effective crisis management (Barton 1993; Coombs 2007; Williams and Treadway 1992; Winsor 1990). Thus far, however, the majority of the research work examining crisis communication has focused on the prevention and recovery stages. Still there is a noticeable dearth of published research that focuses on the communication practice during crisis response. In addition, few studies have been done on crisis communication in the context of the construction industry.

Crisis response is the point which is characterized by short decision time, stress, complexity, and uncertainty. Immediate and appropriate communication decision is vital and even affects the recovery stage management. This research aims to examine the communication practice of Chengdu's construction firms in the critical periods following the Sichuan earthquake crisis.

Moreover, the conventional scientific management of crises encourages a focus on prediction and control, involves "flawless" emergency planning process which includes "a series of checklists or a template" (Thayer 1998, p. 12) for organizations to communicate and respond to the internal and external stakeholders affected. This detailed planning approach often over-simplified the complex by reducing the uncertainty of the situation to a set of rules and steps (Dorner 1996), and presented a limited view of crises as simple calculations involving factors such as crisis type and locus of responsibility (Hearit and Courtright 2004). In fact, these reductionism methods as well as traditional linear communication model did not describe flexible reactions to the real-time changing circumstances that characterize most crises. Hence, there is significant need for a more flexible and adaptive approach of crisis communication management in response to these crises. This research uses the Sichuan earthquake as the backdrop to explain the underlying patterns of communication and decision making of construction firms in response to the earthquake, in order to provide an adaptive approach for crisis response communication management, drawn upon complexity theory.

Furthermore, studies have shown that crisis has been examined by communication researchers from a variety of methodological, theoretical, and structural instances (Benoit 1995; Coombs 2007; Seeger et al. 2001). Although crises take many forms, communication scholars have typically focused on crisis in organizational, financial, or political contexts. In contrast, very limited work has been done on natural crises such as hurricanes, floods, fires, and earthquakes. These events are typically large-scale or geographically based, and their impacts can be severe and highly unpredictable, which require effective response to threats. This research investigates the interests and concerns of the construction firms in response to an earthquake disaster in Sichuan China, and provides a comprehensive understanding of the nature of intraorganizational communication practices in the aftermath situation of a natural disaster.

1.4 Research Aim and Objectives

The aim of this research is to develop a model for effective crisis response communication management, grounded on a conceptual complexity-informed framework. The objectives of the research are to:

1. Explore the nature of the communication practice and challenges during the crisis response stage in construction organizations;

2. Investigate the relevance of complexity theory to the crisis response communication system;
3. Identify significant variables which influence the control level of the communication system for a construction organization in order to be able to creatively respond to unforeseen events and overcome adaptive challenges; and
4. Develop a complexity-informed framework to analyze and explain the underlying pattern of communication, and the behaviors and decisions of the construction organization in response to a crisis.

Based on the above objectives, this study examines the effectiveness and adaptivity of the communication management of a construction organization in response to a crisis. It is recognized that the crisis response communication system within a construction organization should have the characteristics of flexibility and effectiveness, creativity, and adaptivity to adapt to the complex and changing crisis environments.

1.5 Research Hypotheses

Six hypotheses for empirical investigation in this book are formulated below:

H1: The effectiveness and adaptivity of the communication management and decision-making practice of a construction organization in response to a crisis is influenced by organizational structure (OV1).

H2: The effectiveness and adaptivity of the communication management and decision-making practice of a construction organization in response to a crisis is influenced by organizational culture (OV2).

H3: The effectiveness and adaptivity of the communication management and decision-making practice of a construction organization in response to a crisis is influenced by information technology capability and information management system (OV3).

H4: The effectiveness and adaptivity of the communication management and decision-making practice of a construction organization in response to a crisis is influenced by management and leadership style (OV4).

H5: The effectiveness and adaptivity of the communication management and decision-making practice of a construction organization in response to a crisis is influenced by sense-making capability of the organization and members (OV5).

H6: The effectiveness and adaptivity of the communication management and decision-making practice of a construction organization in response to a crisis is influenced by skill capability of members of the organization (OV6).

1.6 Research Scope

This book focuses on internal communication of construction firms, while understanding that this may also be affected by communication with external bodies such as the national agencies and the general public. Successful internal communication on construction jobsites not only reduces the threat to employee health and safety and the loss of corporate or public property, but also stand a better chance of protecting, even enhancing the reputation of construction firms and securing trust from external stakeholders during and/or after the crisis.

The focus of this research is on the communication practices of construction firms in Chengdu city in the critical periods immediately following the Sichuan earthquake of May 12, 2008. It applies complexity theory to analyze and explain the crisis response communication system of a construction firm as a complex adaptive system. How the complex system responds to the earthquake disaster can provide valuable insights into its likely evolution to the next phase in its performance, and its likely actions to prevent recurrence in a disaster environment in order to reduce losses to lives and properties.

1.7 Research Methodology

This research employed a combination of quantitative methods (survey research) and qualitative methods (case study research), with reference to the research objectives stated in Sect. 1.3. This research was conducted in four phases, namely: (i) research model development; (ii) survey instrument development; (iii) case study development; and (iv) research validation, which combined both the qualitative and quantitative approaches.

In the survey instrument development phase, a survey design with the use of a structured questionnaire was adopted. The survey data were collected mainly via a self-administered postal structured questionnaire combined with face-to-face interviews and follow-up telephone discussions with targeted respondents. The survey research identified the significant organizational variables which influence the crisis response communication management process. The results obtained in this phase were used for the case study research in the subsequent phase to gain an in-depth understanding of organizational communication system in response to a crisis. This research study examined the construction firms in the metropolitan area of Chengdu city, the capital of Sichuan province, which is around 80 km away from the earthquake epicenter Wenchuan County in China (See Fig. 1.1 adapted from BBC News (2008)). This massive earthquake has brought severe infrastructure damages and created high levels of threats and challenges for effective crisis response. Records of the Construction Department in Sichuan Province adapted in Fig. 1.2 illustrate some of the infrastructure damage to homes, factories, hospitals, and schools in Sichuan. The questionnaire packages were sent by post or

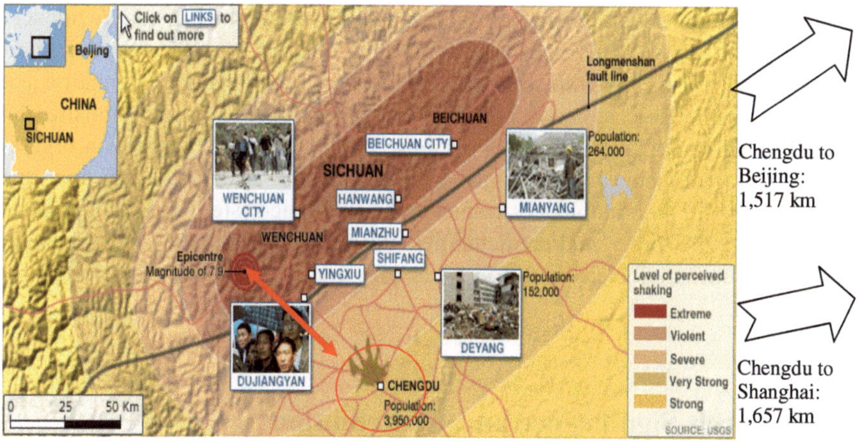

Fig. 1.1 Map of 5/12 Sichuan Earthquake in 2008

Fig. 1.2 Photographs showing the infrastructure damages in Sichuan

personally delivered and targeted to group of contractors and construction firms faced with a major natural crisis during the period of the 5/12 Sichuan earthquake of 2008 in the metropolitan area of Chengdu city, China.

The case study development phase of this study adopted case study/evidence-based research design. The case study data were collected with face-to-face semi-structured interviews and with secondary data source including documentation and archive records. The case study research approach provided a more holistic and in-depth understanding of the crisis response process of the construction organizations, the emergent organizational communication structure, and how the organizations and their employees responded to a chaotic system and communicate about the risk, the threat, and predictability. Within the case studies, two Chengdu construction firms affected by the Sichuan earthquake of 2008 were selected for investigation.

1.8 Research Significance and Contribution

This research would contribute to the knowledge in the following ways:

1. This research would link and extend the knowledge of complexity theory on communication management in the context of crisis response.
2. This study would develop a new mental model to explain and analyze the underlying pattern of communication, behavior, and decision of the construction firms in response to a crisis. It would help to provide a more comprehensive understanding of the nature of organizational communication practice in the situation of a crisis.
3. This study would provide a flexible and adaptive approach for the construction project manager to communicate and respond quickly and effectively in the midst of a crisis. Industry practitioners can use this approach as a planning and evaluation tool for the crisis response communication management.
4. This study would help to enhance the organization's adaptability and resilience in the event of a crisis and contribute to the organization's business continuity management.

1.9 Structure of Book

The remaining chapters are organized as follows:

Chapter 2 presents the main theoretical concepts and ideas underlying the complexity theories and shows how these concepts can be applied to the understanding of social systems and organizational study, especially in the context of the construction industry.

Chapter 3 provides an intensive theoretical review of mainstream communication management and crisis management concepts.

Chapter 4 builds on the concepts presented in Chaps. 2 and 3 to develop a model of crisis response communication management from the perspective of complexity. Findings in organizational studies and management, communication management, and crisis management will be tied back to complexity theory.

Chapter 5 provides the details of the research methodology adopted for this study.

Chapter 6 presents the comprehensive study regarding communication management and practice of the Chengdu construction firms in response to the massive Sichuan earthquake crisis 2008 using the proposed model. Organizational variables which affect the control parameters of the crisis response communication system are identified and examined based on data from fieldwork. Two case studies relating to various aspects of crisis response communication management are also presented.

Chapter 7 summarizes the findings and discusses the implications of complexity theory for crisis communication management and the role of crisis managers in improving the adaptive capability of the organizations. Limitations of this research and suggestions for future research are also provided.

Chapter 2
Complexity Theory Review

2.1 Introduction

As the Nobel Prize winner in chemistry, Ilya Prigogine, wrote: "We believe that we are actually at the beginning of a new scientific era. We are observing the birth of a science that is no longer limited to idealized and simplified situations but reflects the complexity of the real world, a science that views us and our creativity as part of a fundamental trend present at all levels of nature" (Prigogine 1997, p. 7).

Complexity theory is an emerging field of study that has evolved from several major knowledge areas: mathematics, physics, biology, life science, economics, organizational science, and computational intelligence since mid-twentieth century (Holland 1998; Kauffman 1996; Stacey et al. 2000). It is a needed set of endeavors brought about by two realities. The first is that modern science often does not reflect all of reality, but only the part of reality that is ordered, linear, isolatable, predictable, observable, and controllable. The second reality is that modern trends toward disciplinary specialization run counter to the major need for knowledge integration and transdisciplinarity for resolution of contemporary issues.

The idea in complexity theory reverses the traditional views of project management developed over the past 50 years and from a Cartesian/Newtonian paradigm to a more "complex" view (Cooke-Davies et al. 2008). Although there is no universal definition of complexity theory, its principles have inspired many academics and practitioners in the field of business management and provided them with useful explanatory frameworks to understand the behavior of organizations as complex systems (Mitleton-Kelly 2004). Generally, complexity theory is very much concerned with the study of the dynamics of complex adaptive systems (CAS) which are nonlinear, have self-organizing attributes, and emergent properties (McMillan 2006). The underlying idea is that all things tend to self-organize into systems. With complexity theory as the new theoretical framework, this chapter reviews the key findings and core concepts in complexity theory and its applications to the social systems and organizational studies.

Z. Ying and L. Sui Pheng, *Project Communication Management in Complex Environments*, DOI: 10.1007/978-981-4560-64-1_2, © Springer Science+Business Media Singapore 2014

2.2 The Emergence of Complexity Science

2.2.1 *The Origins of Chaos Theory*

Chaos theory or chaos science is the forerunner of complexity theory. The early discovery of the principles of chaos by French mathematician Henri Poincare (1854–1912) is the phenomenon of extreme "sensitivity to initial conditions." In the complex systems of Poincare, small differences grow exponentially with time (Baker and Gollub 1990). Poincare showed mathematically that "small differences in the initial conditions produce very great ones in the final phenomena. A small error in the former will produce an enormous error in the latter. Prediction becomes impossible, and we have a fortuitous phenomenon" (as cited in Crutchfield et al. 1986, p. 48). Two examples were discussed, the first involved the collision of billiard balls and the second involved the weather. Each was ignored by his peers until Edward Lorenz used this theory to explain anomalies that could not be interpreted by the formulations of Newtonian Science (Reulle 1991).

The most well-known example of chaos theory is the "butterfly effect" which reflects the findings of meteorologist Edward Lorenz (1963), one of the pioneers of chaos theory. Lorenz (1963) discovered one important aspect of how nonlinearity affects the weather—the principle of sensitive dependence on initial conditions, that a tiny change in a system's initial state does not inevitably lead to small-scale consequences but that minute change can alter long-term behavior very significantly. Weather prediction was no longer seen as a linear interaction among variables but as a nonlinear complex relationship. Hence, the sensitive dependence on initial conditions theorized by Poincare became the real explanation in Lorenz's model of the weather.

Chaos theory has hitherto impacted a number of disciplines. Bohm (1985), Prigogine and Stengers (1984), etc., have written about chaos and complexity in physics. Lorenz (1963, 1993) has explained the weather modeling based on chaos. Langton (1989), Kauffman (1993), and Maturana (1988) have used chaos science to enhance their understanding of cell behavior and population ecology. Whereas chaos theory arose from mathematical applications to different fields in the physical sciences and biology, complexity theory is wider ranging and is used to describe the behavior over time of complex human and social, as well as natural systems (Jackson 2000).

2.2.2 *Complexity Theory Overview*

Growing from chaos theory, complexity science has seriously challenged long-held views in the scientific community about how the real world works. Complexity theory is a broad theory; it is a related group of concepts and tools that all focus on the effect of interacting parts on the systems as a whole (Carroll and

Burton 2000). Complexity theorists do not conceptualize the world around them as linear and mechanistic or cause and effect. Rather they take a holistic, organic, and nonlinear approach at looking at systems and systems-emergent behavior (Anderson 1999). One commonality in the variety of approaches is an antireductionist view. Complexity theorists believe that there will be lack of understanding of the whole system under study if they simply break down problems or systems to the smallest constitute parts. Rather, the parts (whether they be molecules, genes, agents, or individuals) often have surprising effects on the whole, and behaviors or patterns at the system level may "emerge" from the interactions at the lower levels (Carroll and Burton 2000; Kelly and Allison 1999; Pascale 2000).

Ideas from complexity theory had a substantial impact on various disciplines outside the "hard" sciences, such as biological, chemistry, physical, and mathematics, from where they originated, especially sociology (e.g., Waldrop 1994; Byrne 1998; Urry 2003) and organizational sciences (e.g., Stacey et al. 2000; Stacey 2001; Richardson 2005). The underlying concepts and ideas of complexity science have radical and profound implications for organizations and society as a whole. As an emerging theory with few practical examples, there has been confusion as to the significance of complexity in application as a management tool (Goldspink 2000; Rosenhead 2001). This assertion may well be changing because of the focus on complexity by industry leaders for the twenty-first century corporation (Lewin 1999; Brodbeck 2002).

Principles of complexity theory have inspired many academics and practitioners in the field of business management and provided them with useful explanatory frameworks to understand the behavior of organizations as complex systems (Mitleton-Kelly 2004). Generally, complexity theory is very much concerned with "the study of the dynamics of CAS which are non-linear, have self-organizing attributes and emergent properties (McMillan 2006, p. 25)". The underlying idea is that all things tend to self-organize into systems.

With complexity theory as the new theoretical framework, this chapter reviews the key findings and core concepts in complexity theory, which have challenged the traditional reductionism and determinism-based Newtonian sciences. Table 2.1 shows the major concepts and emerging diagram of complexity science and some of the key researchers associated with them.

2.3 The Core Concepts of Complexity Theory

2.3.1 Complex Adaptive Systems

The most important concept introduced by complexity science is the CAS, as defined by Holland (1996), which is presently also commonly denoted as a multiagent system. The basic components of a CAS are called agents. They are

Table 2.1 Summary of main concepts and emerging paradigms of complexity science

Time period	Concepts/paradigms	Key researcher	Discipline
1960s and 1970s	Butterfly effect	Edward Lorenz	Life Science
	Strange attractors	Edward Lorenz David Ruelle	Mathematics
	Self-organizations, Dissipative structures	Ilya Prigogine Stuart Kauffman	Physical Science Life Science
	Self-organizations, Evolution and complexity		
1980s	Edge of chaos	Chris Langton	Life Science
1990s and onwards	CAS	John Holland	Mathematics
	Emergence	Murray Gell-Mann	Physical Science
	Complex responsive process of relating	Chris Langton Ralph Stacey	Life Science Life Science

typically conceived as "black box" systems, meaning that the rules that govern their individual behavior are known, but their internal structure is unknown. The rules they follow can be very simple or relatively complex. Intuitively, agents can be conceived as autonomous individuals who try to achieve some personal goal or value by acting upon their environment. But an agent does not need to exhibit intelligence or any specific mental quality, since agents can represent systems as diverse as people, ants, cells or molecules (Heylighen et al. 2006).

Stacey (1996, p. 10) defines CAS as consisting of "a number of components, or agents, that interact with each other according to sets of rules that require them to examine and respond to each other's behavior in order to improve their behavior." These interactions need not be physical; they may also relate to sharing information (Cilliers 1998) and the interactions develop patterns that are created when a number of simple rules are applied over many iterations (Kelly 1999). The results of individual interactions are unpredictable: small differences at the start of the process can eventually result in large differences in the system's performance. Many interactions in a system can produce unexpected patterns or behaviors (Goldberg and Markoczy 2000) because stimulating one part of the system can have unexpected effects in other, unanticipated, parts of the system. Such unexpectedness is because of the nature of nonlinear feedback networks and the interconnected and interdependent nature of CASs (Stacey 1995).

Although there is no universally accepted paradigm for describing CASs (Gell-Mann 1994), Stacey (2001) summarized the structure of a CAS as follows:

- The system comprises large numbers of individual elements or agents.
- Agents interact with a number of others according to the rules for interaction being "Local", i.e., no individual agent has complete knowledge about the behavior of the system as a whole, there is no system-wide rules determining the interactions.
- These interactions are iterative, recursive, and self-referential, i.e., effects of the interactions are looped.

- The agents adapt to each other and the interactions are nonlinear, i.e., the results of the interactions are unpredictable.
- Rule variety results from random mutation and cross-over replication.

CAS models afford new opportunities for analyzing complex systems without abstracting away their interdependencies and nonlinear interaction. They typically show how complex outcomes flow from simple schemata and depend on the way in which agents are interconnected (Anderson 1999). This is particularly important for organizational scholars because interdependency is central to modern conceptions of what an organization is (Thompson 1967).

The components in a CAS are nested: each individual component is itself a complex system of interconnected parts, and they interact with each other to exchange resources and information necessary to maintain the system's internal balance. To self-regulate, a CAS uses control processes such as feedback loops that move the system toward its internal goal (Capra 1996). The feedback mechanisms require the components of adaptive systems to learn and use their *learning* as the basis of action. There are two kinds of *learning* in CAS (Stacey 2000): (1) single-loop learning when one learns from the consequences of previous actions to amend the next action. It is a feedback process from action to consequence to subsequent action, without questioning the mental model driving the action. This kind of single-loop learning is depicted in Fig. 2.1.

In contrast, (2) double-loop learning, which involves another feedback loop in which not only the actions are amended, but also the model driving the actions. This double-loop learning is depicted in Fig. 2.2.

In both learning processes, the system adjusts its behavior to the stimulus in its environment so that its behavior can be better adapted to the environment; but double-loop learning is much more profound which also involves *creation* and *innovation*—new ways of viewing the world.

Fig. 2.1 Single-loop learning

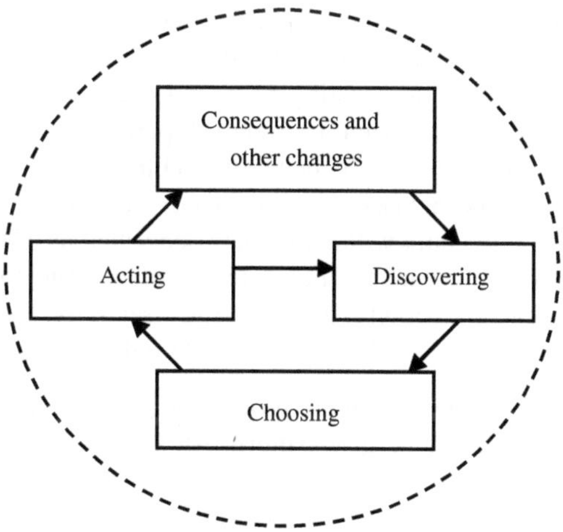

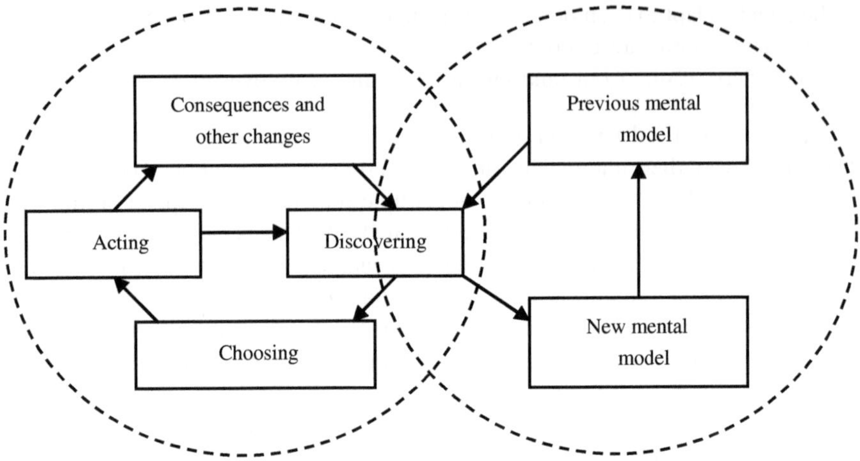

Fig. 2.2 Double-loop learning

Creativity in CAS occurs at the edge of chaos—a space of novelty. Researchers studying CAS have shown that the space for creativity in such systems can be characterized by the following (Stacey 1996):

- A phase transition. The space for creativity in an adaptive system is a phase transition at the edge of chaos, at the edge of system disintegration.
- A state of paradox. Bounded instability is an essentially paradoxical space of simultaneous stability and instability. Examples of this paradox are that information both flows freely and is retained; Agents or components are richly but not too richly interconnected; behavior is both predictable and unpredictable; both order and disorder exist; both competition and cooperation exist.
- Actualization of archetypes. Archetypes are rules that drive the behavior of agents in a system. The actualization of archetypes is sensitively dependent upon the precise interactive experience of the system.
- Creative destruction. In the phase transition, creative change is made possible by destruction. Indeed the double-loop learning process involves destruction of some of the existing rules and creation of new ones.
- A critical point for control parameters. There are three control parameters driving the CASs behavior: (1) the rate of information flow through the system; (2) the richness of connectivity between agents in the system; and (3) the level of diversity within and between schemas of the agents. These control parameters dictate whether the system will migrate to the edge of chaos and remain as endless variety, novelty and creativity, or conversely fall into true chaos—patternless instability and internally unconstrained.

The outcome of the process of learning in a complex system is self-organization producing emergent pattern. All CAS self-organize and create emergent structure that can be better adapted to the changes in their environments. The next section describes these characteristics in more detail.

2.3.2 Self-Organizing Systems

Complexity science is concerned with systems that have the capacity to spontaneously self-organize themselves into even greater states of complexity (McMillan 2006). The physical chemist, Ilya Prigogine, developed the theory of "dissipative structures" which was the first description of what is also called "self-organizing systems" or "self-organization systems" (Capra 1996). According to Prigogine, dissipative structures not only maintain themselves in a stable state far from equilibrium, but may even evolve. When the flow of matter and energy increases, they may go through new instabilities and transform themselves into new structures of increased complexity (Quoted by Capra 1996, p. 89). Self-organization is "the spontaneous emergence of new structures and new forms of behavior in open systems far from equilibrium, characterized by internal feedback loops and described mathematically by nonlinear equations" (Quoted by Capra 1996, p. 85).

Kauffman (1996) describes self-organization as a force of "anti-chaos", noting that some very disorganized systems spontaneously bifurcate and then solidify at a new, higher level of order. Moreover, Kauffman (1996) argues that this new order arises from inner guidelines and principles rather external forces. These chaotic, complex systems have some inner drive or pull toward order, although the relationship between order and chaos is always in dynamic tension. Kauffman (1996) also suggests that this self-organization phenomenon is a quasi-evolutionary adaptive force, leading systems to higher levels of complexity and order following the collapse or bifurcation of a lower order system.

Spontaneity is an important feature of self-organizing systems as they interact and reshape themselves. These ideas apply to shoals of fish, ant colonies, and human social groups. What all these systems have in common is that their ability to spontaneously self-organize, they exchange matter and energy, and remain far from equilibrium. Feedback loops contained within the system ensure that rich patterns are produced and the system itself behaves in its own unique way (Cooke-Davies et al. 2008). Self-organizing systems are complex dynamical systems that appear capable of self-organization and exercising choice in a manner that makes them inherently unpredictable. Self-organization is considered as the spontaneous reallocation of energy and action to achieve a collective goal in a changing environment (Kauffman 1993; Comfort 1994).

From a self-organizing systems perspective, structure emerges from randomness; systems produce order from the edge of chaos (Holland 1995; Waldrop 1994). In the context of construction projects, self-organization may occur all along the project life cycle. For instance, when a project organization's existing structures and standard communication processes are not effective when an unforeseen event or a construction crisis interrupts the flow of the activities and/or disrupts the process.

2.3.3 Emergence

Another important concept is the notion of emergence or emergent properties. The concept of emergence is a main concept that flows through studies of complexity. It is a phenomenon of the process of evolving, of adapting, and transforming spontaneously and intuitively to changing circumstances and finding new ways of being (McMillan 2006). It is at the heart of the process of evolving, adapting, and transforming. Auyang (1998) suggests three criteria for emergence characteristics in CAS. First, an emergent character of a whole is not the sum of the characters of its parts; second, an emergent character is of a type totally different from the character types of the constituents; third, emergent characters are not deducible or predictable from the behaviors of the constituents investigated separately.

Chris Langton (Waldrop 1994) at the Santa Fe Institute described how a global order emerges from the interaction in a local, dynamical system, and in doing so a whole new set of properties emerge. The emergence can be seen as unpredictable patterns of orders that appear through a process of self-organization. Varela (1995), for example, asserted that emergence refers to the point at which a system's local interactions become global patterns encompassing all individual agents. It can be seen in the properties of ecosystems, insect swarms, human societies, and all the other CAS.

It is these emergent properties of living systems that allow novelty and innovation and provides a credible account of how diversity and variety arise in order to allow evolution to happen. It also suggests that when dealing with complex dynamic systems, there is en element of unpredictability about the future that is pregnant with as-yet undreamt of possibilities (Cooke-Davies et al. 2008). As Kauffman (2000, p. 139) puts it, "The universe in its persistent becoming is richer than all our dreaming." The feature of emergence can also be used to explain that how the system will behave cannot be determined by studying its parts. No matter how complete our familiarity with its components, a complex system cannot predict with certainty how or in what direction the system as a whole will develop (Boje 2000).

2.4 The Paradigm of Complexity Theory

The term "paradigm" has become widespread in discussions of the philosophy of science since it was introduced by Kuhn in 1962 in *The Structure of Scientific Revolutions* as a way of describing achievements that arise when a group of scientists adopt models from which spring "particular coherent traditions of scientific research" (Kuhn 1996, p. 10). In the following, an overview of the most significant components of the emerging paradigm of complexity theory will be provided.

2.4.1 *The Butterfly Effects: Sensitive Dependence on Initial Conditions*

During the late 1960s and early 1970s, a group of scientists working in a wide range of disciplines became uneasy with the basic assumptions of linearity that were used as the basis for much science, especially in the physical sciences. It was while using a computer to simulate weather systems in 1960 at the Massachusetts Institute of Technology that meteorologist Edward Lorenz (1963) discovered one important aspect of how nonlinearity affects the weather—principle of "sensitive dependence on initial conditions" (Cooke-Davies et al. 2008). When searching for a means to provide accurate weather forecasts, Lorenz looked into the computer's output for patterns or rules that could be used in forecasting methods. Lorenz found that small difference in the input can generate completely different output. The findings of how minute changes can have major and unpredictable consequences in nonlinear systems became now commonly known as the "butterfly effect" (Gleick 1987).

Holland (1995) identified the nonlinearity as one of the properties common to all complex systems. The nonlinearity of a complex system suggests that a given action can lead to several possible outcomes, some of which are disproportionate in size to the action itself. Holland (1998) also studied the nonlinear agent-based models and specified the mechanisms through which entities, or agents, relate to each other and how these mechanisms form the building blocks of the model. The widespread presence of such nonlinearity is seen to make prediction impossible over large swathes of the natural and social science (Jackson 2000).

Paradoxically, the fact that complex dynamic systems, such as the weather system studied by Lorenz, are not predicable in the long term does not mean that it is impossible to understand or even to explain their behaviors. The system may appear to be behaving erratically and unpredictably at first glance, but observation over a longer time period or on a wider visual scale will show patterns emerging that echo each other and weave around to form an unexpectedly stable tapestry of behaviors (McMillan 2006). Lorenz's weather model indeed exhibited a fine geometrical structure, some kind of order in the middle of chaotic systems (Gleick 1987). This butterfly effect challenges traditional ideas of cause and effect and concept of predictability.

2.4.2 *Strange Attractors*

Attractors are well-known behaviors of nonlinear complex systems. The most famous attractor is the Lorenz's strange attractor, which was another contribution of Lorenz's (1993). To understand strange attractors, it is necessary to know that in any dynamic complex system (e.g., a simple pendulum), one can represent the state of the system using a diagram known as phase space. In phase space, the system operates within a basin of attraction. This figurative basin is "where the systems explore millions of possibilities, wandering to different places. But its wandering and experimentation respect a hidden boundary which is gradually revealed as the shape of its strange attractor (Wheatley 2006, p. 118)." The system does not wander off into infinity, the boundary lives within the system.

An adaptation of the Lorenz's Strange Attractor in Fig. 2.3 depicts the chaotic behavior of a complex system with a three-dimensional graphical representation (with the arrows denoting dynamism in a complex environment). It shows the space where the point ends up after many iterations. As the system's chaotic wanderings are plotted over time, the attractor reveals itself. Although the specific path of behavior in chaos is unpredictable, that behavior does have a hidden pattern, a qualitative shape (Stacey 2000). Hence, the butterfly strange attractor reveals the order inherent in a chaotic system (Gleick 1987).

The importance of the Lorenz attractor resides in the fact that it allows for reducing the space of states in which one can find the system after a few interactions. This insight discovers that complex systems can follow a number of qualitatively different attractors, depending upon initial conditions and external perturbations—that is very different from simple deterministic chaos (Cooke-Davies et al. 2008). To see how chaotic processes reveal the order inherent in a system requires that we shift our vision from the parts to the whole (Wheatley 2006).

Stacey (1996) further defines an attractor as "a pattern of behavior into which a system ultimately settles in the absence of outside disturbances (p. 54)." There are three kinds of attractors: stable, unstable, and strange. Both stable and unstable attractors are well-known behaviors of deterministic nonlinear feedback system. A stable attractor uses negative feedback to decrease small, gradual disturbances, pushing a system back to equilibrium. A stable attractor generates a predictable

Fig. 2.3 The Lorenz's strange attractor

pattern. An unstable attractor uses positive feedback to amplify disturbances, pushing a system toward complete instability and randomness. Only a strange attractor is simultaneously stable and unstable, for example Lorenz's butterfly attractor which exhibits a general pattern, although the specific position of any point cannot be predicted.

2.4.3 Edge of Chaos

Perhaps the most intriguing finding of the complexity theory is that many systems undergo a phase transition between order and disorder, between stability and chaos (Packard 1988; Langton 1992; Kauffman 1993, 1996) that has been called the "edge of chaos", a term featured prominently on the whole emerging field of complex theory from the life sciences. Studies of the evolution and behavior of living dynamical systems suggest that these systems manage to demonstrate elements of both chaotic and orderly behavior, and both computer scientists and evolutionary biologists have carried out pioneering work to understand why this might be the cause (Lewin 1993).

Kauffman (1996) uses water as an analogy to describe the edge of chaos. Water exists as solid ice, liquid water, and as gaseous steam. Kauffman's hypothesis is that if any living system become too embedded or too deeply involved in the frozen, highly ordered area (e.g., ice), then it becomes too rigid to undertake the complex activities necessary to sustain and develop life. On the other hand, if the system becomes too embedded in the gaseous chaotic zone (e.g., steam) then it would suffer from a complete lack of order and again would be unable to carry out all the activities necessary to survive. Thus, the best place for a living system to exist is in the fluid area (e.g., water) which lies between the two other areas.

Research on CAS has found that they have the ability to exist and operate in a state that is between pure stability and complete instability in a region that contains both stability and instability. A central characteristic of CAS is that they are precariously poised on the edge of chaos, which is a certain kind of balance between the forces of order and disorder (Kaufman 2006). Those systems in the ordered regime have a tendency to "freeze" into a fixed state. Here minor alterations are unlikely to have major effects on the system. There is order, but little adaptation. In contrast, disordered systems rarely settle down into stable patterns. Instead, small changes cascade though the systems, causing multiple other small changes as each element alters those it is connected to.

The place where stability and instability occur simultaneously was dubbed "the edge of chaos" by Langton (1992), founder of "artificial life". A complex system must operate in a far-from-equilibrium condition, analogous to the edge of chaos, since nothing novel can emerge from systems with high degrees of stability (Horgan 1996).

2.4.4 Complex Responsive Process of Relating

Complex responsive process of relating (CRPR) is a theoretical concept within the conceptual palette and CAS in particular, which has been introduced and argued for by Stacey (2000, 2001) and other co-researchers (Stacey, Griffin and Shaw 2000; Griffin 2002; Streatfield 2001) on the basis of the problematic capacity of other theoretical approaches to address complexity in contemporary organizations. It suggests a particular way of speaking about complexity of organizations, organizing, managing, and knowing. It emphasizes the reflexive nature of humans, the essentially responsive and participative nature of human processes of relating, and the radical unpredictability of their evolution and outcomes over time (Cooke-Davies et al. 2008).

The attention of previous sections on the complexity theory is drawn to self-organizing, emergent property of the system, the dynamics of the edge of chaos, as well as the possibility of unpredictability. However, all of these analyses tend to be at the macro level of the organization as a whole in system terms. Stacey's (2000) well-known statement on CRPR explores further to emphasize at a micro level and concentrate on the dynamics of bounded instability in which self-organization might produce emergent novel forms in relating and conversation.

From a methodological point of view, this concept puts emphasis on the interaction among people in organizations and is concerned with the question of how patterned themes of conversations in local situations constitute and are simultaneously constituted by power relations in organizations, and how the potential transformation of these conversational patterns can induce change, trigger learning, and create new knowledge. With these assumptions and propositions, CRPR implies an alternative view on management of organizational arrangements, the methodology of inquiry, the possibility of control, and the role of individual and the group in these processes (Cooke-Davies et al. 2008).

Focusing attention on the responsive process of relating between people encourages a different way of thinking about organizations and the nature of control. Organizations, no matter how large, are processes not things. They are continuously reproduced and transformed in the ongoing communicative interaction between people, both their formal members and people in other organizations (Stacey 2001). It includes three essential aspects for organizations' business process: (1) the processes are processes of *relating*. That is to say that all processes are always related to some other actors, e.g., other processes. These other actors may be within the same organization or in other organizations; (2) the processes *respond* to the needs and expectations from the related other actors; and (3) the processes are always *complex* because all details of expectations and interests cannot be in full certainty defined or agreed. Hence, all the processes include always some degree of risks on not fulfilling all needs and expectations. Figure 2.4, adapted from Stacey (2001), describes different categories of process based on the degree of certainty and level of agreement. From a management point of view, there is not one simple management approach which is sufficient to

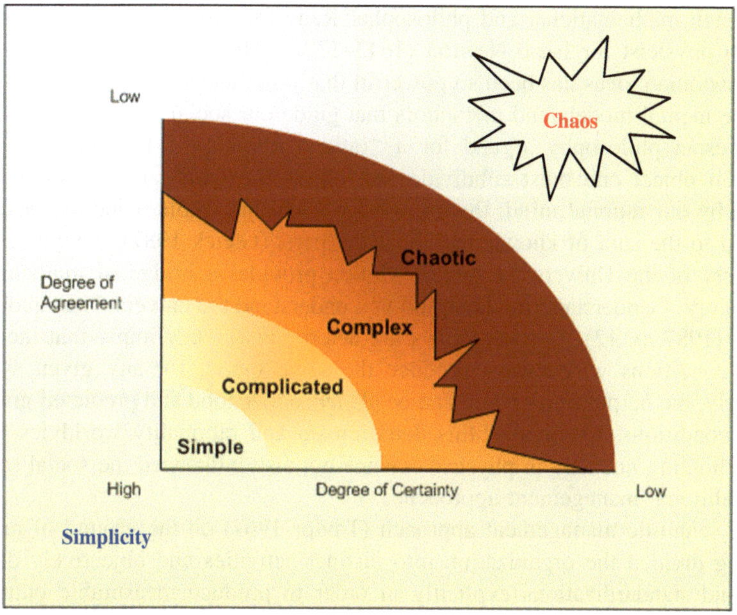

Fig. 2.4 Agreement versus certainty matrix

manage all differing features of a business process; instead, appropriate actions should be selected based on the certainty and agreement on the issue in question.

From the perspective of CRPR, organization is an emergent property of many individual human beings interacting together, centered on the use of language simultaneously for conversation and to negotiate social status and power relationships. The interactions can be general conversations, technical discussion, formal letters, emails, telephone calls, etc., anything that conveys an idea, opinion, etc. Communication is a complex process of relating—a chain of patterned responses that provide the context for an individual action across space and over time (Stacey 2000). Communication by means of evolved language is a defining characteristic of human beings, distinguishing them from other species of animal (Kauffman 1993). Central to the theory is the recognition that communication is a complex process involving both the words that are spoken and the response that they elicit.

2.5 Classical Scientific Perspective Versus Complexity Perspective

Over centuries, worldviews and paradigms have influenced science and common beliefs within society as a whole. Since the seventeenth century, Western worldview has been influenced by two important figures in philosophy and science:

the French mathematician and philosopher Rene Descartes (1596–1650) and the English physicist Sir Isaac Newton (1643–1727). The combination of Cartesian and Newtonian ideas has been so powerful that it has shaped not only science but also the mental models and metaphors that guide our social behavior as well.

Cartesian philosophy argued for a "rational thinking": To gain knowledge about an object one must subdivide that object into parts small enough to be known by our rational mind; the knowledge about the whole object is, therefore, reduced to the sum of knowledge about the parts (Gleick 1987). Newton's great discovery of the Universal Laws of Motion provides a bridge of mathematical uncertainty. "Understand the laws and you understand the universe", as quoted by Gleick (1987, p. 12). Newton's laws are deterministic: they imply that the same initial conditions will always produce the same output for any given system. Similarly, the output of a system can be always understood and predicted given its initial conditions. Eventually, this determinism and rationality worldview influenced thinking not only in physical science but also influenced the social science and traditional management approaches.

The scientific management approach (Taylor 1967) on the ground of rational thinking divided the organization into distinct activities and objectively defined rules and standardizations explicitly in order to produce measurable output by creating a linear working environment. As observed by Wheatley (2006, p. 6), "We manage by separating things into parts, we believe that influence occurs as a direct result of force exerted from one person to another, we engage in complex planning for a world that we keep expecting to be predictable, and we search continually for better methods of objectively perceiving the world".

Although the scientific management approach may be an effective management tool within a relatively stable and simple organizational environment, it is an unrealistic approach to deal with today's increasingly complex and dynamic business environment. In particular, the notion of planning and control affects the flexibility of organization and reduces the system adaptability. This is a server limitation when it comes to encounter the unforeseen changes or unexpected situations such as crisis. Table 2.2 provides a comparative perspective of a number of key stereotypical approaches and patterns of thinking between Newtonian-Cartesian world view and complexity science world view (Carlisle and McMillan 2002; Capra 1996).

Hence management of complex dynamic systems requires a new framework that abandons the traditional linear thinking patterns and cause and effect models, and instead applies the new nonlinear approaches and the recognition that linear approaches are inherently over simplistic. The next section presents how the complexity theory can be applied to the contemporary social system and organizational studies.

Table 2.2 A comparative perspective of Newtonian-Cartesian and complexity science worldview

Newtonian-Cartesian perspective	Complexity science perspective
Essentially mechanistic	Essentially dynamic/self-organizing
Linear	Nonlinear
Controllable	Uncontrollable
Centralized	Networked
Hierarchical	Nonhierarchical
Limited connectivity	Highly connected
Uniformity	Diversity
Cause and effect	Effect and effect
Predictable	Unpredictable
Reductionist	Holistic
Objective focus	Subjective and objective foci
Entity focused	Process focused
Correlation	Patterning
Highly preclusive	Highly inclusive
Evolutionary	Revolutionary and evolutionary

(*Source* adapted from Carlisle and McMillan (2002) and Capra (1996))

2.6 Complexity Theory and Its Application to Social System and Organization

Within the scientific domain, complexity science has successfully challenged the traditional western scientific worldview as discussed before. It also introduces innovative and even radical thinking into a number of nonscientific domain such as economics (Arthur et al. 1997; Krugman 1996; Ormerod 1999), the social sciences (Allen et al. 2003; Byrne 1998; Gilchrist 2000), and organizational management (McMillan 2006; Stacey 1996, 2001; Wheatley 2006).

2.6.1 Development of Organization Theory

The notion of the organization as an orderly machine with a hierarchical structure is one that has been both widespread and potent for many years (McMillan 2006). Max Weber (1864–1920), a German sociologist who is credited as a major contributor to the development of the theory of bureaucracy, asserted that hierarchy, authority, and bureaucracy are at the roots of all social organizations (Pugh and Hickson 1996). Weber's model considered that the bureaucratic form of organization is a powerful tool in that it creates an environment that encourage people to respect and follow the rules, thus making them more efficient. Hence, the rational use of authority and control was highly desirable for an organization. The value of this bureaucratic model highly relies on its predictability and top-down control.

Pugh and Hickson (1996) also pointed out that Weber's model had a set of rules and procedures which aimed to cover every possible situation that might arise within the organization, which again presented a worldview that believes that events can be predicted and planned for with a large degree of certainty.

By the 1950s and 1960s, the systems approach to organization emerged, taking a more holistic approach to organizations and building on the notion that organizations, like organisms, are open to their environment (McMillan 2006). Priesmeyer (1992) attributed this approach to a developing awareness of the organization's relationship with its environment. It was a move to recognize the human behavioral factor but it still promoted the notions of authority and control and the chain of command.

The prevalent literature on organizations and organization theory has been challenged in recent years by a number of writers. Handy (1994) in his noted *The Empty Raincoat* presents new and challenging visions of reconsidering life in organizations. Morgan (1997) uses metaphor to explore our notions of organizations in an exciting and transforming way. Senge (1990) and Pedler (1991) develop a set of useful ideas on learning and the learning organization.

More recently, a number of researchers are now looking to the new science of complexity for ways of better understanding the structure and the dynamics of modern organizations with the present turbulent and demanding time. Englehardt and Simmons (2002) suggested that an organization can be perceived as a systemized whole that comprises many interdependent and coordinating components. Moffat (2003) highlighted that complexity theory provides an explanatory framework of interrelationship of how individuals and organizations interact, relate, and evolve within their environment. Moffat (2003) highlighted that this theory attempts to explain why interventions may have unanticipated consequences and also explore the effect of these consequences on the relationship between interactive elements.

Pascale et al. (2000) offer a new management model derived from complexity science which is based on four core principles, on the recognition of the powerful influence that science has on society. These are as follows (Pascale et al. 2000):

(1) "Equilibrium is a precursor to death." When a living system is in a stable state then it is not readily responsive to changes and so it is placing its survival at risk.
(2) When faced with either an opportunity or a threat, living systems move toward the edge of chaos. This is because here they are able to mutate and experiment and so fresh solutions to these challenges are more likely to be discovered.
(3) When living systems do this, they undergo a self-organizing process and new forms and new behaviors emerge from the upheaval.
(4) Living systems cannot be directed in a linear fashion as unpredictable consequences are inevitable. The challenge for organizations is to disturb these systems in a way that is similar to the outcomes desired.

Another influential researcher Ralph Stacey (1996, 2001) demonstrated the contribution that notions from complexity apply literally to organizations and business and are not simply to be considered as useful analogies or metaphors. In Stacey's view, if managers want to adopt a scientific approach to management, they need to understand the behaviors of nonlinear feedback systems in unstable or far-from-equilibrium situations. This is because, organizations are just such systems. Lewin and Regine (1999) also support Stacey's view that organizations are CAS as they have all the properties of these systems. An organization is a place in which self-organizing relating between people in which power, politics, and conflict of ordinary, everyday life are at the center of co-operative and competitive organizational processes through which joint action is taken (Stacey 2000).

2.6.2 Construction Organizations as Complex Adaptive Systems

A CAS in the natural domain consists of a very large number of agents which interact with each other and together form a system that adapts to its environment (Kauffman 1996). This agent-based approach is of particular interest to social/management scientists because human groups, organizations, and societies may also be thought of as agents interacting with simple rules because of their sociological and psychological nature. From the complexity standpoint, organizations are CAS comprised of agents (people) who experiment, explore, self-organize, learn, and adapt to changes in their environment. They exist at the individual, team, divisional, and group level and also in a much larger web of external CAS—their economic, social, and political environments.

Stacey (1996) emphasized the importance of learning as a feature of CAS and points out that human systems are adaptive because they are engaged in double-loop rather than single-loop learning. A self-organizing sand pile does not learn and adapt, but a human self-organizing team does. Hence, it is learning that differentiate CAS apart from other complex and complicated systems (McMillan 2006).

Another important feature of CAS is that they have emergent properties. These systems are able to create emergent outcomes through a process of spontaneous self-organization. In nature, an organization emerges from self-organizing processes in the form of organism that co-evolve with multiple others, as CAS. The self-organizing nature of communicative interaction among organizational members, i.e., their joint action, is always "contextually" mediated by the participating individuals making reference in their conversations to symbols and artifacts representing the "situational rationality" (including formal structures, procedures, plans, contract document, etc.), which are in turn reflections of the patterns of routines and power relations in the process of organizing (Weick 1995; Stacey 2001).

On the other hand, human organization emerges not only from self-organizing process, but also from human interventions (Espejo 2006). Hence, as Stacey (1996) noted, outcomes are partly determined by self-organizing agents and partly by intentional choice of management. Organizing forms need to reflect these forces. Stacey (2001) presents an understanding of "organization" as an emergent property of many individual people interacting together through their complex responsive processes of relating, centered around the role of language that is simultaneously used for conversation and to negotiate social status and power relationships.

Lewin and Regine (1999) consider organizations as CAS and organizations can learn a lot about their own dynamics by understanding and learning all about these systems. To successfully survive in today's fast-changing business environment, organizations will need to behave as CAS operating on the edge of chaos. They also provide a large number of examples of organizations which have managed to find ways of achieving this.

A construction project organization incorporates parts of several organizations, each a subset of the interests of its own organization. The project teams are linked by various types of contractual relationships and professional rules that determine the organization structure and project delivery system. The multiplicity of decision makers from different organizations with different goals and values influence resource allocation within shared problems oriented to the achievement of a plan or end result (Wild 2002). Hence, construction project organization is a multi-agents system in which diverse agents interact with each other in a nonlinear way and exchange resources necessary to achieve project goals.

In this multiple-agents system, each agent barely, if ever, is aware of all factors influencing the system. Instead, each agent operates on the assumptions of what other agents in the systems are going to do. Human (agent), by nature, construct internal models to deal with uncertainty (Arthur 1994). Based on the past experience and limited knowledge, construction agents predict and construct an internal models of others' behavior to deal with the uncertainty of the situation. They select from the most creditable models and try them out. If the outcomes are positive, the models are reinforced and polished, whereas, if the results are negative, the models are impaired and eventually displaced with a new model. Moreover, the construction agents should also reflect upon their project environment and flexibly reorganize their project team so that it can better respond to the uncertainty and changes of the environment such as a crisis. By the capability of double-loop learning (Stacey 1996) from the past experience, agents are continually adapting and co-evolving within a complex environment (Arthur 1994).

It has been argued that construction organization should be understood as a complex, dynamic system. It can be been seen as a social system—a cooperation between individuals and groups brought together for the project (Tavistock 1966). The group working on the project consists of the individuals, and the team spirit and cooperation are emergent phenomena. However, practice in construction unfortunately often shows that the lack of cooperation makes the whole very inefficient compared to the sum of the participants, and the cooperation has a hard time to emerge. Instead, self-interest and group-interest grow as another kind of emergent phenomena (Bertelsen 2003).

Looking at construction as a social system, the groups and individuals brought together for the purpose of executing the project cannot be considered completely autonomous, but they are to a great extent equal individuals as they formally belong to other organizations than the one established for the project execution. As the project progresses an informal project organization emerges along with the formal organization established by the project management. Meanwhile, the team members create an informal communication network that is superimposed on existing formal project structures. This informal communication networks are emergent outcomes which are not planned by a project organization and thus cannot be predicted.

Considering construction organization as a CAS challenges our traditional definition of control and command. The conventional definition of control indicates the central authority and a powerful tool of bringing the system back to its desired, initially planned stable state. Conversely, in CAS, control is distributed among the system's agents, it reflects the agents' ability to self-control and regulate. Professionals in a project organization can share the control of the project, and be actively involved in the decision-making process. As complex systems are in their nature unpredictable but capable of self-organization and learning, management of such systems cannot be based on detailed planning but must comprise a statement of the objective, improvement of reliability and distributed control (Bertelsen 2003).

2.7 Summary

The study of complexity theory has migrated from mathematics and physics through biology and ecology toward the soft science, e.g., economics, social science, and organizational management. The ideas contained in complexity theory define a number of challenges for traditional project management while at the same time offering a potential solution to the unacceptably high rate of failures. Complexity theory, as linked to social system and organization studies in this chapter, suggests that human organizations are CAS comprised of agents (people) who experiment, explore, self-organize, learn, and adapt to changes in their environment. The idea of further extending complexity theory to construction organization management has also been investigated. It is proposed that a construction project organization can be considered as a CAS operating in an uncertain and complex environment. Faced with the internal and external discontinuities, the organization self-organizes, allows new informal communication patterns to emerge and adapts to the dynamically changing situations.

Armed with these basic principles of complexity theory and possibility for their application to social systems and in particular construction project organization, it can be then moved on to address a complexity-based or complexity-informed framework for communication management during a crisis. The next chapter presents a comprehensive theoretical review of mainstream crisis management and communication management.

Chapter 3
Theoretical Review on Crisis Management and Communication Management

3.1 Introduction

Crisis can happen to any organizations and crisis management is crucial for all organizations because effective crisis management helps to ensure the continuous well-being of an organization. The importance of crisis management is highlighted if viewed in the light of the nine knowledge areas in the Project Management Body of Knowledge (PMBOK) by the Project Management Institute. Communication is particularly challenging during various stages of crises, and effective communication management is a critical tool in the management of a crisis situation. It plays an importance role in how an organization resolves a crisis it cannot avoid.

Therefore, the question arises—How should construction companies manage the communication process and overcome communication challenges in response to a crisis? To answer that, the concept of mainstream crisis management and communication management is reviewed in this chapter.

3.2 Crisis and Crisis Management

3.2.1 Definition of Crisis

Before discussing the management of crises, it is important to understand the definition of what constitutes a crisis. In layman terms, the Oxford Dictionary defines a crisis as "a time of intense difficulty or danger; a time when a difficult or important decision must be made; a turning point in the course of anything" (Oxford Dictionaries 2011). In a general term, a crisis is "a situation faced by an individual, a group or an organization, which they are unable to cope with, by the use of normal routine procedures and, in which stress is created by sudden change" (Booth 1993, p. 86). Fearn-Banks (2001) explained that a crisis is a major occurrence with a potentially negative outcome affecting the organization,

Z. Ying and L. Sui Pheng, *Project Communication Management in Complex Environments*,
DOI: 10.1007/978-981-4560-64-1_3, © Springer Science+Business Media Singapore 2014

company, or industry, as well as its publics, products, services, or good name, and can sometimes threaten the existence of the organization.

The term "crisis" has been defined differently by various management writers and the particular factors emphasized in a given definition of crisis therefore differ accordingly. In terms of the causes to a crisis, Lerbinger (1997) organized crises into seven different types: (1) natural, (2) technological, (3) crises of confrontation, (4) crises of malevolence, (5) skewed management values, (6) deception, and (7) management misconduct. These cause-related definitions encourage a view of crisis in terms of events impinging from the outside, to which the organization then reacts (Gilpin and Murphy 2008).

Some researchers have defined crises by their psychological factors, particularly from a manager's point of view. Lerbinger (1997) described three psychological issues when the managers face a crisis: (1) extreme time pressures to act; (2) lack of clarity about what is the best action to take; and (3) an element of surprise. Weick (1993) described the psychological impact of crisis as a cosmology episode that jolts an individual's entire belief system. Similarly Pearson and Clair (1998) highlight the subjective perception of organizational crisis: "An organizational crisis is a low-probability, high-impact situation that is perceived by critical stakeholders to threaten the viability of the organization and that is subjectively experienced by these individuals as personally and socially threatening" (p. 66). They further extended the construct of crisis to consider facets of psychological, social-political, and technological-structural perspective jointly.

Other authors have favored crisis definitions that focus on its effect on the members of the organization and other important stakeholders. Many authors agreed that a situation becomes a crisis when one or more stakeholder groups perceive it as such (Gilpin and Murphy 2008). A rumor, or simply the perception of a crisis event, is sufficient to trigger a response, regardless of whether the rumor or perception is grounded in facts (Coombs 2007). This definition of crisis is perceptual in that it is the perceptions of stakeholders that help to define an effect as a crisis.

Although crisis is often perceived to be a sign of managerial failure, there is also a duality view of crisis for both danger and opportunity (Fink 1986). Bronn and Olson (1999) defined crises as the product of "either a threat or opportunity that arises from internal or external issues that may have a major impact on an organization" (p. 355). Fink (1986) and Health (1998) also described crisis as a period of sudden change during which a totally new system is formed; stressing on the fact that the meaning of crisis does not only cover risk, uncertainty, threat, conflict, accident, and instability but also covers opportunity.

The above discussion of various definitions of organizational crises indicates that the managerial attitudes and behaviors and decision makings before, during, and after crises can be crucial and closely linked to the continuous well-being of an organization. Section 3.2.2 will examine the mainstream view of crisis management.

3.2.2 *Crisis Management*

3.2.2.1 Overview of Crisis Management

The field of crisis management has evolved from the relatively long tradition of research into disaster management (Shrivastava 1993). Crisis management represents a set of factors designed to combat crises and to lessen the actual damage inflicted (Coombs 2007). Gigliotti and Ronald (1991) defined crisis management as the ability of an organization to deal quickly, efficiently, and effectively with contingency operations with the goal of reducing the threat to human health and safety, the loss of public or corporate property, and adverse impact on continued normal business or operations.

A typical crisis requires large amounts of information because little is known from the start. It is a rapidly changing situation, and usually changes are difficult to predict. What distinguishes a crisis from a day-to-day problem is the extreme sense of urgency that hyper-extends an organization's coping capabilities, producing stress and anxiety among organizational members and other stakeholders (Allen 2001; Pearson and Clair 1998). From this perspective, the highly reactive emphasis of crisis management can be seen as opposed to the proactive emphasis of the fundamental term in project management—risk management.

The term risk according to the PMI (2004) is an uncertain event or condition that, if it occurs, has a positive or a negative effect on a project objective (such as time, cost, or quality). A risk may have one or more causes and, if it occurs, one or more impacts. The most common risk management technique used is classification of potential causes, identification of potential risks, and the assessment of mitigation strategies according to risk profile. Risk assessment and management is a well-established knowledge area in modern project management. In theory, risk events are probabilistic, are able to occur regardless of the likelihood, and therefore each risk should be responded to within the project framework in some way. The greater focus of risk management is on risk mitigation, where potential threats are highlighted and solutions are put in place to avoid these risks.

However, when a project risk of critical consequences has not been successfully mitigated against, a crisis occurs. The term "crisis" is a general glossary that can be described as, for example "a situation faced by an individual, a group or an organization, which they are unable to cope with, by the use of normal routine procedures and, in which stress is created by sudden change (Booth 1993, p. 86)". When a crisis will undoubtedly cause a significant disruption to an organization, a business continuity strategy and plan can help minimize the disruption. Research shows that organizational contributory factors affect the tendency of executives to adopt an effective "crisis as opportunity" mindset (Brockner 2008). James (2010) contends that most executives focus on crisis management as a reactive strategy. While the company's reputation with shareholders, financial well-being, and survival are all at stake, potential damage to reputation can result from the actual management of the crisis issue. Crisis leadership, on the other hand, immediately addresses both the damage and implications for the company's present and future conditions, as well as opportunities for improvement (James and Lynn 2007).

By its very nature, crisis management is multidisciplinary (Fink et al. 1971; Mitroff 1988; Shrivastava 1993), which unites management theory with psychological, social-political, and public relations perspective to create a comprehensive model of crisis process. Researches on crisis management examined the development of crises over stages, crisis planning, crisis decision making, and the communication strategies available to organization in the midst and aftermath of a crisis (Allen and Caillouet 1994; Benoit 1997; Coombs 1995; Fink 1986).

Coombs (2007) considers crisis management as a set of four interrelated factors: (1) prevention, (2) preparation, (3) response, and (4) revision. *Prevention*, also known as mitigation, represents the steps taken to avoid crises. Crisis mangers often detect warning signs and then take actions designed to prevent the crisis. *Preparation* includes the crisis management plan (CMP) and involves diagnosing crisis vulnerabilities, selecting and training a crisis management team, and refining a crisis communication system. *Response* is the application and testing of the preparation components to a crisis. An organization's crisis management response is frequently reported and critiqued in the news media (Pearson and Clair 1998). Finally *revision* involves the evaluation of the organization's response in simulated and real crises, and revising its prevention, preparation, and response efforts.

The early crisis management literature has focused primarily on the development of CMP (Coombs 2007). This field was concerned mainly with tactical advice that prescribed specific plans and checklists. A CMP "consists of a full range of thoughtful processes and steps that anticipate the complex nature of crises real and perceived" (Caywood and Stocker 1993, p. 411). Because CMPs are developed in anticipation of crises before they occur, CMPs allow for speedier and more efficient responses to crises when they do occur (Barton 1993, 2001; Coombs 2007). One component of the CMP is the development of a strategy for what to say in the wake of a crisis (Coombs 2007). What organizational representatives say to organizational members and other stakeholders after a crisis has occurred is referred to by Coombs (1995) as a crisis response strategy. Crisis response strategies are message repertoires that are designed to repair the organization's image by influencing stakeholder perceptions.

Increasingly, scholars and practitioners have begun to give more attention to strategic issues, focusing on areas such as issues management and environment scanning, noting the impact of contingency and uncertainty (Millar and Health 2004; Mitroff and Anagnos 2001). For example, Fearn-Banks (2007) defined crisis management as "a process of strategic planning for a crisis or negative turning point" (p. 2). More organizations now have strategic planning than ever before, and that number continues to rise (Coombs 2007; Ofori 1994). Most recently, crisis management studies have started to emphasize the culture drivers of crisis and the social construction of crisis (Gilpin and Murphy 2008). One recent trend in crisis management also focuses on development OF strong, positive relationships with various stakeholder groups and the interaction between internal and external drivers of crisis (Coombs 2007).

In fact, today's environment seems to be placing higher premiums on crisis management, and unprepared organizations have more to lose today than they ever have before (Coombs 2007). Organizations are playing for high stakes when confronting crises; hence there is in great demand for organizations to improve their crisis management process. Section 3.2.2.2 discusses the different models for the crisis management process and approach.

3.2.2.2 Crisis Management Process and Approach

Crisis management is a dynamic and continuous process that includes both proactive and reactive actions with the aim of identifying the crisis, planning a response to the crisis, confronting the crisis, and resolving the crisis (Ocal et al. 2006). Studies in crisis management indicate clearly that the phenomenon of crisis is articulated from different phases or stages (Low et al. 1999). The early four-stage model by Fink (1986) is used as a medical illness metaphor that a crisis can consist of as many as four different and distinct stages: (1) prodromal or hints of a potential crisis begin to emerge; (2) acute, a triggering effect occurs; (3) chronic, the effects of the crisis linger as efforts to clean up the crisis progress; and (4) resolution, there is some clear signal that the crisis is no longer a concern to stakeholders.

Mitroff (1994) provided a more prescriptive five-stage crisis management process model with concern on how crisis management efforts progress. It is concluded that regardless of the type of crisis, effective crisis management involves managing the five distinct phases through which all crises pass: (1) signal detection: new crisis warning signs should be identified and acted upon to prevent a crisis; (2) preparation and prevention: organization members search known crisis risk factors and endeavor to reduce their potential for harm; (3) damage containment: a crisis hits and organization members try to prevent the crisis damage; (4) recovery: organization members work to restore normal business operations as soon as possible; and (5) learning: review and critique the crisis management efforts, and revision. The failure to manage any one of these phases well may be responsible for the occurrence of a crisis in the first place and then for its escalation.

A more general three-stage model of crisis management process which has accommodated the other two models has been recommended by a variety of crisis management experts (Birch 1994; Coombs 2007; Seeger et al. 2003): pre-crisis, crisis, and post-crisis. Richardson (1994) provided the first detailed discussion of these three macro stages: (1) pre-crisis stage, encompasses all the aspects of crisis preparation, including prodromal signs, signal detection, and prevention, where warning signs appear and people try to eliminate the risk; (2) crisis or crisis impact stage, includes damage containment, recovery or chronic sub-stage, where the crisis hits and actions and support is provided for those involved in it; and (3) post-crisis, includes the learning and resolution sub-stage, where the crisis is resolved.

Overall, an appreciation of the different phases of crisis management helps crisis managers to better understand the complexity and nature of crisis management (Coombs 2007). Effective crisis management is vital for all the organizations.

3.2.2.3 Stakeholder Theory in a Crisis Context

Stakeholder theory (Freeman 1984) is primarily concerned with how groups and individuals affect an organization and managerial behavior taken in response to those groups and individuals (Frooman 1999). A stakeholder is defined as any group or public affected by the organization's operations (Ray 1999). The nature of the relationship between the stakeholder and the organization is important in shaping the response to stakeholder pressures (Stephens et al. 2005).

A crisis can be the perception of an unpredictable event that threatens important expectancies of stakeholders and can seriously impact on organization's performance and generate negative outcomes (Coombs 2007). Crises may disturb some stakeholder expectations resulting in people becoming upset and angry, which threatens the relationship between the organization and its stakeholders. Crisis studies have now begun to emphasize the interaction between internal and external multiple stakeholders. Development of strong, positive relationships with various stakeholder groups as a preventive measure or attenuating factor in the event of a crisis has been focused on by several crisis management scholars (Caponigro 2000; Coombs 2000; Zhong and Low 2007b).

In response to a crisis, organizations need to recognize a broad number of their stakeholders including organization members, public, media press, etc. When an organization's environment is complex and unstable, internal and external stakeholders cross over boundaries as they become involved in the crisis (Leibinger 1997). Horsley and Barker (2002) made suggestions within their public agency model which predicts greater success of crisis management if information is disseminated quickly, accurately and candidly to critical stakeholders, including the media. Indeed, Irvine and Millar (1996) found that the vast majority of organizational crisis were not the result of technical failures or environmental damage but were instead the direct result of the organization's inability to develop and maintain positive relationships with key internal and external stakeholders.

The emphasis on the critical relationships makes stakeholder theory particularly useful to stakeholder management in the situation of crisis. For instance, Heath (1998) stressed the importance of developing strong pre-crisis relationships with stakeholders, arguing that organizations should focus on building mutually beneficial relationships with stakeholders and focus on an appropriate sense of corporate responsibility in their pre-crisis, mid-crisis, and post-crisis communication. Although establishing strong stakeholder relationship will not likely help an organization avoid or avert every crisis, it can play an important role in how the organization resolves a crisis it cannot avoid.

3.2.3 Crisis Management in Construction

3.2.3.1 Types of Construction Crises

Preparing for the unexpected situation is crucial in the construction industry because it may happen every day. As indicated by statistics relating to the construction industry, a realistic definition of a crisis in construction typically means that someone has been injured or killed on a project (Reid 2000). Construction risks and crises have primarily been identified based on the causes of the crises, in order to prevent them in the future. Table 3.1 provides a list of potential crises in construction (Perry and Hayes 1985; Reid 2000).

Differing in the timing, Jarman and Kouzmin (1990) distinguished between creeping, periodic, and sudden crises. Creeping crises are generally systemic; they are often something that should have been anticipated and seen as inevitable at some time or others. A periodic crisis can be the impact of business cycles, economic cycles, and other changes that flow in a predictable way but where the timing is not easy to accurately predict. An example of speculative building in the housing sector can often lead to the developers being subject to financial crises. A sudden crisis is one that occurs seemingly from nowhere and often appears overwhelming. Sudden crises are usually the result of exceptional contributing factors. Natural disaster falls into this category, for example, unseasonable and extraordinary rainfall may undermine a foundation wall. Other examples include a fatality of a workman or collapse of temporary support wall caused by unusually high wind conditions.

Typical analysis of construction and real estate crisis is based on economic, legal/regulatory, and political aspects. For instance, the sudden collapse of Asian economies during the 1997 financial crisis has been the subject of studies for construction and real estate companies, but most of these studies focus on economic fundamentals (Lu and So 2005). The results of a study by Ocal et al. (2006) indicated that governmental policy and unstable market conditions mainly caused the 2001 economic crisis for Turkish construction companies. Kaklauskas et al. (2011) pointed out the effect of the general economic crisis on the business relations toward increasing distrust between the market members in the construction industry.

Crises come in different sizes (Kapucu and Van Wart 2006) and result in different level of damage, injury, commotion, disruption, hostility, or media attention the event provokes (Reid 2000). Generally speaking, the names given to the smallest emergencies or crises are hazards, incidents, or simply emergencies. They can come from day-to-day routine construction processes. The expectation is that the response will be handled entirely at the local level.

Large-sized crises are often called disasters and are events that cause considerable loss of life and/or property damage (Kapucu and Van Wart 2006). A disaster is defined by the Asian Disaster Reduction Center (2003) as "A serious disruption of the functioning of society, causing widespread human, material or

Table 3.1 Potential construction crises

Category	Example
Natural disaster	Earthquake
	Lightning
	Extreme snow/ice conditions
	Extended freeze
	Flood/drought
	Landslip
	Hurricane/tornado/tsunami
Operations	Equipment failure
	Accident involving a company vehicle
	Loss of a key subcontractor/supplier
	Construction delay
	Cost overrun
	Design error/issue
	Explosion
	Fire
	Major utility failure
	Neighborhood/community opposition to a project
	Structure/subsidence collapse
	Data/telecommunications failure/loss of critical data
	Fluctuations in market demand for product or service
Environmental accidents/liabilities	Groundwater contamination
	Long-term exposure of the community to toxic chemicals
	Ecological damage and pollution
	Public inquiry
Management issues	Bankruptcy
	Contractual dispute with a client, resulting in litigation
	Negative publicity from rumors
	Availability of funds
	Serious cash flow problems
	Sudden market shift
	Lack of specialized resources
	Loss or damage in the transportation of materials and equipment
Government affairs	Legislation that could affect business
	Political risks in countries of owners and/or suppliers and/or contractors
Employee safety and health	Chronic safety problems
	Exposure to carcinogens
	Injury/fatality of an employee or nonemployee
	Personal injury suit
	Regulatory citations
Employee/management misconduct	Disgruntled employee
	Executive misconduct/fraud/embezzlement
	Price fixing
	Sabotage
	Theft/vandalism
	Workplace violence
	Scandal involving top management

(*Source* adapted from Perry and Hayes (1985) and Reid (2000)) environmental losses which exceed the ability of affected society to cope using only its own resources (available at www.csao.org/images/pfiles/278_Emergency ResponsePlanning)". There are many research centers and agencies all over the world that are concerned with disaster management (prevention, preparedness, mitigation, response, and relief) (Shaluf 2007a). It has been found that disasters can be classified into three types: natural; man-made; and hybrid (Shaluf 2007b). Natural disasters are catastrophic events resulting from natural causes, such as volcanic eruptions, tornados, earthquakes. In contrast, man-made events refer to technological failures or malicious incidents. Power outages and computer-related attacks, such as hacking, and viruses are prevalent examples.

The term crisis and disaster are often used synonymously. Although disasters come from a class of crisis, there are some technical differences between the terms emergency and disaster. Crisis/emergencies are often local in nature and response, described as "any incident that can focus negative attention on a company and have an adverse effect on its overall financial conditions (Reid 2000, p. 2)," "those internal and/or external events that cause stress on organizational resources and pose the greatest treats on any organization's security and vitality (Hallgren and Wilson 2008, p. 831)." More specifically, a crisis/emergency is an unexpected event locally in an organization's life which demands a time-pressured response.

However disasters have a broader geographic scale and complexity. They may be handled entirely at the local level, but they often require a regional or national response with multiple agencies providing direct assistance. For instance, Comfort et al. (2001) suggested that when disaster threatens a community, it requires inter-organizational coordination to focus time, effort, and attention on a common goal. In particular, natural disaster crises present major damage to human health and environment and result in loss of human life, physical, and financial damages. Disaster environments create an extraordinarily difficult context for crisis coordination for the construction organizations. The distinction is most important when state or local disaster declarations fund recovery and remediation efforts. Also in the post-crisis of a disaster such as that of an earthquake, the communication infrastructure and network may be fully/partially destroyed/damaged, thus rendering communication more difficult. On the contrary, this is not usually the case in the crises that come from the day-to-day construction processes where the existing communication infrastructure and networks are usually intact, i.e., not damaged.

It is of extreme importance for construction firms to provide an immediate, adept crisis response to protect lives, health, and the environment. Here in this study, the terms crisis/emergency and disaster will be used as defined above. This study provides analysis of the internal communication practice of construction companies in a natural disaster—the Sichuan earthquake in May 2008, and how the project managers in Chengdu communicate and respond to the employees following the Sichuan earthquake.

3.2.3.2 Importance and Challenges of Crisis Management in Construction

As for all other organizations, effective crisis management is equally important for construction organizations. Construction is a complex, crisis-prone activity carried out in an environment which is relatively uncontrollable (Galbraith 1973). All construction companies are prone to crisis but this varies depending on the nature of the work they are engaged in (Walker and Loosemore 2003). Crises do not discriminate—small construction companies or large, specialized or general (Reid 2000), and seem inevitable in projects (Hallgren and Wilson 2008). Construction companies that deal in projects on an ongoing basis thus must learn to deal with crises on a regular basis and take prudent steps in anticipation of their occurrence.

Efforts in construction management research have usually been directed toward the search for improved predictive and preventative techniques (Sinclair and Haines 1993). However, Loosemore (1998a) highlighted that in construction there are important differences between the projects based temporary multi-organizations used to procure buildings and the relatively permanent organizations within which the majority of crisis management research has occurred. In particular, project organizations are typically more dynamic and characterized by higher differentiation and conflict, which make the management of change more difficult. A further reason for caution in transferring general crisis management principles to the construction context is related to the construction industry's unique professional roles, employment practices, expectations, norms, and traditions. Hence it is important to recognize the possible influence of these unique features on the crisis management process.

It is impossible to eliminate the possibility for crises in any organizations. A crisis-free environment is unlikely to be achieved in any organizational context. Of course, this is especially true for a construction organization which is exposed to relatively higher risk of technical accidents and uncertainty. Evidence suggests that many construction organizations exist in a low state of crisis preparedness, having an inadequate understanding of their risk exposure, of how to mitigate those risks and of the internal systems needed to cope with, learn from and recover from their eventuality (Teo 1998).

Effective crisis management can prevent loss of money, limit reputation damage, and reduce the time it takes to complete the crisis life cycle. Moreover, if the organization could utilize communication management more effectively in the event of the crisis, it would be more prepared to resolve the crisis which cannot be avoided. In Sect. 3.3 the mainstream communication management is examined.

3.3 Communication Management

Communication is both a process and an activity. It is a process of information exchange using a common system of symbols, signs, or behaviors (Cleland 1994; PMBOK 2004). It is an activity that consists of defining the communication needs and expectations for the project; how, when, in what format and by whom and to whom, information will be exchanged; it is based on the requirements of the stakeholders. Successful communication is not only developing the plan, but also implementing the plan for continuous engagement with stakeholders (PMBOK 2004).

3.3.1 Theoretic Models of Communication

The most common application of communication theory is in understanding the transfer of information in the form of acoustic or visual messages, the function of which is to convey meaning (Dainty et al. 2006). Shannon and Weaver (1949) developed a simple linear model of communication supported by mathematical theory, which has relevance to the ways in which people communicate, interpret, and disseminate information. The early linear model of communication such as Shannon's (1948) and Minai's (1984) has traditionally dominated research within and outside a construction management context (Rogers and Kincaid 1981). This simple model [See Fig. 3.1 adapted from Emmitt and Gorse (2003)] comprises a transmitter relaying information as a signal to a receiver, the efficacy of which is affected by noise, which distorts the clarity of the message to the receiver, who subsequently decodes the message. Every transmission has to be encoded in order for it to be communicated because otherwise the message could not propagate (Skyttner 1998). Noise is important part of this process as it can impair the transmission of the message from the sender to the receiver, and cause any type of distortion or distraction that can affect the quality of transmission between parties.

However, the early linear model has been criticized for masking a far more complex set of parameters that combine to shape the way in which information is transferred and interpreted (Dainty et al. 2006; Emmitt and Gorse 2003). For example, it ignores the fact that all communication is potentially a two-way process, with receiver providing feedback message to the sender on whether the message is understood. Also it does not consider the physical and social context in which the communication parties work. Thus, although it provides a good starting point for understanding communication process, it fails to satisfactorily convey the process of human interaction.

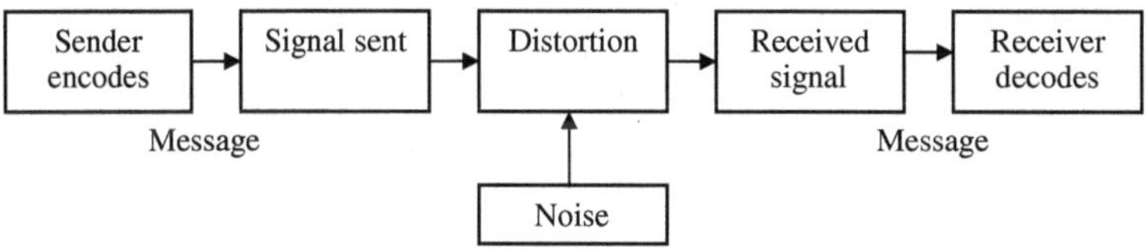

Fig. 3.1 Linear model of communication process

Fig. 3.2 Two-way model of communication process

A more sophisticated model which portrays communication as a two-way process has emerged to produce a more comprehensive theory (Baguley 1994; Feldberg 1975; PMBOK 2004). The application of communication theory in this model [See Fig. 3.2 adapted from Baguler (1994) and PMBOK (2004)] has focused on the effects of messages on the receiver with particular emphasis on "feedback" that was used by the sender to improve and adjust their messages. The sender creates the message, encodes and sends it to the receiver; the receiver interprets the message, decodes and creates a feedback message to acknowledge the sender that the message has been received and understood. Finally the sender receives the feedback, decodes and checks that the receiver has understood the message as intended.

This model also includes reference to the communication medium and channels along which it passes. Viewing communication as a process in this way presents it as a more dynamic or iterative concept where the transmitter is continually receiving feedback and adjusting his/her decision. The aim of this two-way communication is to persuade the "receivers" or "audiences" about something perceived to be of value or interest to the "senders." It involves a strong element of persuasion and control by the sender (Foster and Jonker 2005). "Noise" exists in every communication medium and distracts the receiver from the contents of messages. A well-designed message is clear and easy to understand by the receivers. According to Crane and Liversey (2003), this communication model places the focus on the information itself as a commodity that needed to be transmitted rather than seeing communication as a social process that brings meaning to life through negotiation and consensus (Smircich and Stubbart 1985).

Adding the dimension of context within which the communication process takes place in the form of structures, cultures, group task characteristics, and information from the environment, Thompson and McHugh (2002) provided a more advanced model, presenting an open system view of the communication process. Figure 3.3 [adapted from Thompson and McHugh (2002) and Dainty et al. (2006)] depicted this model considering the construction industry context within which communication takes place.

Despite the increasing sophistication of theoretic models discussed above in presenting the communication process, they all have weakness in that they viewed

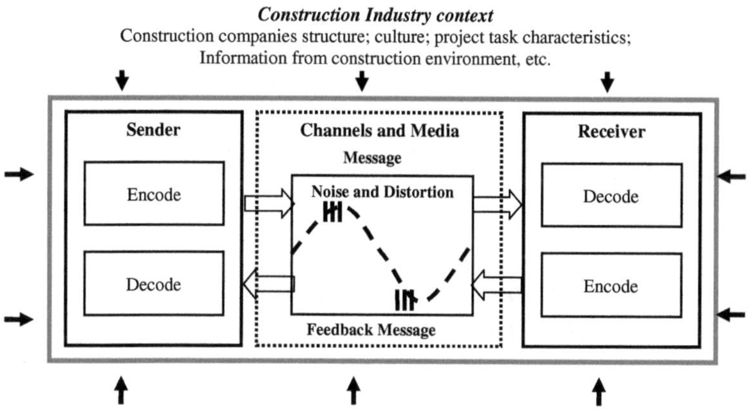

Fig. 3.3 Communication process for construction industry

communication as a sequential rather than a simultaneous or concurrent process (Thompson and McHugh 2002, p. 261). In these models, communication is framed as a flow, a transmission of information and orders occurring within an organizational container constructed by managers. Because the container determines the form of transmission, these models deem it logical for managers to reduce the cost and variability of communication in the interest of control and command. This perspective ignores the interactive dynamics of communication with characterizes most form of human interaction (Dainty et al. 2006).

These conventional models of communication made assumptions that the communicator could control the message in the sense that it could determine how it was perceived by the audience. Krippendorff (1989, p. 71) criticized that "most communication models are one-way in the sense that they start with a sender and end with the effects on a receiver and thereby equate communication with control." Grunig (1992) called these "asymmetrical dialogue" where the aim of the communication is to manipulate or persuade, even though it may involve two-way interaction. Indeed, these communication models mask the complex interplay of signs, meanings, and symbols which are often more important than the explicit communication itself (Fiske 1990). Although these models helped to illustrate what was being processed and the mediating organizational factors which affected the communication performance, they viewed communication as a sequential rather than a simultaneous or concurrent process. They also did not explain and recognize the interactive dynamics and complexity of communication which characterized most forms of human interaction.

In practice human communication is dependent on cognitive ability (Emmitt and Gorse 2003). An understanding of cognition is essential to understand how communication is taking place, for which a background of shared social reality should exist. The message is not passively received and understood. Rather, the receivers or audiences actively develop meaning, and this is created in terms of their perspectives on the world in which they live and the concrete situation at hand (Foster and Jonker 2005). For example, the architect and construction manger need to have an inclination of what the other person might understand to communicate effectively. Also the same facts and observations may manifest in two different people; however, this does not mean that they make the same assumptions, although they may be capable of doing so. Assumption about each other's knowledge and experience needs to be made.

3.3.2 Internal and External Organizational Communication Contexts

Communication can be regarded as the substance of everyday organizational life (Eisenberg and Goodall 1993, p. 18). It is an essential aspect of the functioning of an organization as well as its information exchanges with its environment (Rogers and Agarwala-Rogers 1976, p. 7). Thus organizational communication can be considered in either internally or externally defined terms (Dainty et al. 2006):

- The internal dimension focuses on ensuring effective communication between managers and employees and among employees in different parts of an organization. This is vital for regulating employee behavior in a way that helps to deliver on the organization's objective, for innovation in changing the way things are done, and information needed by employees to perform in their duties.
- The external or inter-organizational communication dimension focuses on information exchange with external parties, such as clients, local communities, trade unions, and national agencies.

Although these two perspectives to some extent represent different kinds of processes, internal and external communication should be seen as mutually intertwined, as if one is ineffective, then the other is likely to be detrimentally affected. Thus, a failure of either communication process is likely to impact detrimentally on the organization's performance (Emmitt and Gorse 2003). Understanding communication within organizations is vital because without effective communication systems and procedures, they cannot manage the complex flow of information necessary for interaction with their internal and external environment. Accordingly, the study of communication in organizations usually examines the flow of information through channels and networks and the contents of messages sent (Thompson and McHugh 2002, p. 260). This presents a particular problem within the construction project environment where the distributed and temporal nature of project teams undermine direct control and influence over the way in which information is interpreted and managed.

Organizational communication can also be distinct in formal and informal forms of interaction. Formal communications are the accepted system of communication within the organizations; they are the official sources of information using prescribed channels. Informal channels are routes of communication other than those identified by the organizations, e.g., emerge through friendships or contacts between individuals who are willing to cooperate (Emmitt and Gorse 2003). By their nature, formal communications tend to be more authoritative than informal communication, but may also be structured to hide information as well as communicate the selected information (Weaver 2007).

3.3.3 Communication and Stakeholder Management

Regular, two-way communication between an organization and its internal and external stakeholders is the life blood of a favorable organizational-stakeholder relationship. A number of scholars (Bendell 2000; Crane and Liversey 2003; Foster and Jonker, 2005) have suggested that the essential building-block of stakeholder relationships is communication. Crane and Liversey (2003) suggest that stakeholder relationships nowadays are characterized by a complex array of shifting, ambiguous, and contested interactions between interested parties and within diverse organizations. This highlights the central role of communication in constituting, managing, and maintaining stakeholder relationships.

According to Harrison and St John (1996), stakeholder management includes communicating, negotiating and contracting, managing relationships, and motivating them to respond to the organization in ways that benefit it. One widely applicable approach is for an organization to use communication or open dialogue with their stakeholders (Freeman 1984; Harrison and St John 1996; Polonsky 1995). Deetz (1995) suggested that the stakeholder model can enable responsible practices when complemented by adequate conceptions of communication. Certainly, for an organization to be truly responsible to its stakeholders, it must engage in communicative processes that enable complete and open representation for all the parties.

According to Bakhtin (1981), the process of dialogue is one that strives to incorporate diverse voices while developing synergy, empathy, and authentic deliberation with a relationship. Deetz (1995) extended this definition and referred to dialogic communication as that in which meaning is always incomplete and partial, and the reason to communicate is to better understand all parties of the relationship and ultimately find new and satisfying ways of interacting. Deetz (1995) suggested that most corporations use forums for interaction that suppress or diffuse inputs from stakeholder groups rather than fostering genuine dialogue with them. Forums for dialogs and representation should be considered in assessing any form of responsibility. A dialogic perspective entails communicative processes that encourage honest engagement in which values, assumptions, and the needs of others are openly discussed and addressed. Waddock (2001) suggested that valuing and respecting others in the stakeholder relationships, as well as valuing the relationship itself is at the heart of dialogue. Dialogue offers a way to communicate that respects all parties and involves mutually interactive and interdependent relationship building.

Unfortunately, Liedtka (1998) suggested that dialogic processes, while important to stakeholder relationships and social responsibility, are generally not fully implemented by organizations. Waddock (2001) argued for organizations to engage with stakeholders on an interactive basis and to employ a dialogue-based approach. Waddock further observed that this is not easy for organizations, nor do many organizations yet engage with organizations interactively. This is often difficult as stakeholders differ in their interests as well as their values to the organization.

A vital component of building and maintaining relationships, communication is essential for maintaining the support of commitment of all stakeholders (Briner et al. 1996). Effective, regular, planned, and ad hoc communication with all members of the project community are necessary for project success (Cleland 1994). A project manager must be able to recognize the danger signals, the warning of possible trouble with senior stakeholders. These danger signals take the form of actions such as interfering in the business of the project without consultation, not providing support when needed, poor communication links caused by too many reporting levels between the project manager and the senior stakeholder, unfounded promises or commitment (Boddy and Buchanan 1999).

3.3.4 Perspective of Organization and Communication

Communication is the process within which social reality is constructed (Krippendorff 1989). All social systems are necessarily communication systems (Luhmann 1995) and humans "live in communication" (Pearce 1989:196). To say that humans live in communication is to assert that process per se is formative; what is formed is the shape of our interactions and our meanings about them. According to Milar (2003), communication, as an evolutionary process, is ongoing, imperfectable, relational, and emergent.

In the early 1980s, organizational communication researchers began to express dissatisfaction with the traditional perspective on their object of inquiry, arguing that information-based conceptions of communication (Shannon and Weaver 1949) failed to provide an understanding of the activity of organizing, and that such views encouraged people to ignore the dynamic nature of organizational structure. For instance, Smith's (1993, 1996) overviews of conceptions of organization and communication showed that the literature is replete with a container metaphor, in which the organization is seen as a container, or vessel, and communication is simply a phenomenon that occurs inside its walls. Axley (1996) argued that most management literature, as well as management studies, conceive of communication with a conduit metaphor, a tool that facilitates transfer of messages from point to point.

Smith (1996) and Axley (1996) argued for the conception of a meaning-based view of communication, in which the important process is not the generation or transmission of messages, although they should not be neglected, but rather is the meaning and ways that communicators jointly construct through their use of symbols, which is dynamic over time. Fulk (1993, p. 924) stated that though there are considerable differences between such perspectives, they usually share the core proposition that social and symbolic processes produce patterns of shared cognitions and behaviors that arise from forces well beyond the demands of the straightforward task of information processing. Indeed, the importance of communication processes is the unfolding relationships between actors, which can be considered as an ongoing and continually involving interaction. Communication processes are not "contained" within organizations, but are the essence of how people organize (Farace, Monge and Russell 1977). In other words, organizations are built as people act in patterned ways and as those patterns become externalized and made to seem objective (Berger and Luckmann 1966). Therefore, communication is inherently organizational and organizing is always communicative.

Taylor et al. (1996) approached the notion of a duality of structure to develop a theory of the equivalence of communication and organization. "We believe the organization is not in the activities as such, but in their interpretation. And the working through of an interpretation is a social process (in fact the process we call communication) by means of which members both come to an understanding of what the events mean, organizationally, while they simultaneously reconfirm their own position in the network through the role they play in the interpretive process. We therefore need to show how the organization emerges in the communication in two ways, both as a more or less share understanding, and as dynamic playing out of the relationships where identifies and roles get negotiated" (Taylor et al. 1996, p. 3).

The organization then is seen as a portable carrier of socio-technical knowledge, skills, and procedures of making sense of the world, through which particular patterns of behavior and action emerge and reproduce themselves in specific material and social circumstances via human communication (Reed 1996). Chia (2002) considered that organizations should not be understood as stabilized objects, but as a spatial–temporal framework "for institutionalizing social habits and patterns of behavior so that it then becomes possible for us to communicate with each other and develop practical norms" (Chia 2002, p. 867).

Hence, an organization is an information processing and communication system, structured to achieve a specific set of tasks and goals. Rather than describing a physical entity, the term "organization" in this report is meant to designate the process of organizing which signifies systems as "ongoing activities that are accomplished mainly through communication" (Hawes 1990).

3.4 Communication Management in Crises

3.4.1 Crisis Communication Management Overview

Scholars have long recognized the important role of communication in effective crisis management (Barton 1993; Coombs 2007; Millar and Heath 2003; Winsor 1990). For example, Coombs's (2007) authoritative book on crises management focused almost exclusively on communication issues. Winsor (1990) documented how communication failures in the warning stage led to the Space Shuttle Challenger accident and showed how communication breakdowns in the warning stage can actually exacerbate a crisis situation. Marlow and Wilson (1997) considered that effective communication management is a critical tool in the management of a crisis situation.

When an organization is threatened by external environment crises (e.g., natural disaster) or internal events (e.g., structural change), the need for communication increases to some extent. The role of the communication manager or project manager as part of top management and strategic decision making is becoming increasing important (Grunig 1992). "Excellent" organizations use the potential of communication management to assist in transformation and relationships with the environment (Grunig 1992). Communication management can turn around the negative effects caused by small events that result in chaos.

In the context of construction organizations, communication difficulties occur because magnified conflicts of interests increase the importance of information as a source of power in negotiations and make people more secretive with it (Loosemore 2000). Ocal et al. (2006) concluded that ineffective communication management is one of the most important factors affecting the decisions and the implementation of the decisions, in their examination of the construction organizations during the economic crisis in 2001.

As mentioned in Sect. 3.4, the crisis management process can be divided into three stages: pre-crisis, crisis, and post-crisis. Appropriate crisis communication strategies and decisions are taken corresponding to different crisis stage. Section 3.4.2 discusses the crisis communication during the crisis response stage, which is also the research focus in this book.

3.4.2 Crisis Communication: Response Stage

Once a crisis hits, communication presents unique challenges during the response phase. Internally the project managers must collect and process information to make decisions and communicate with the members properly. Externally, stakeholders must be informed and communicated about the crisis situation and appropriate crisis response communication strategies must be taken to address it (Coombs 2007).

3.4.2.1 Conventional Crisis Response Communication Model

The crisis response stage is entered when avoidance efforts fail and events trigger a crisis. At this stage, organizations shift their resources and efforts to minimize damage to the people, facility, and environment. Conventional crisis management planning process provides "a series of checklists or a template" (Thayer 1998, p. 12) which can help the organization go into auto-pilot; guide managers how to classify and prioritize crisis events, and how to communicate information tailored to different key stakeholders, including employees, clients, government authorities, media, press, etc. (Barton 2001; Coombs 2007; Mitroff and Pearson 1993).

Crisis response communication normally includes conveying ongoing crisis events to stakeholders, decision making with the crisis management team, and organizational decisions regarding whether and what amount of information to share (Hale et al. 2005). Through intensive interviews with crisis decision makers and detailed reviews of secondary data sources, Hale et al. (2005) concluded that the typical crisis response communication practices usually involve four interdependent procedures (Hale et al. 2005). Figure 3.4 [adapted from Hale et al. (2005)] illustrates two conventional models—(a) linear and (b) cyclic models used to describe how organizations communicate during the crisis response stage.

The linear communication model follows a pattern of sequential steps. Upon a triggering event, examined crisis response communication patterns can be described as initiated by *observation* which entails assessing and gathering relevant information, followed by *interpretation* which involves understanding and analyzing the information, *decision* which includes examining, discussing, filtering, and choosing alternatives among crisis decision makers, and concluding with *dissemination* which involves implementing the decisions, distributing, and exchanging the information with key stakeholders.

In terms of the cyclic model, iterations of the four process steps several times are conveyed. It attempts to provide a better description of the complexities of communication processes following the triggering of a crisis, by continually assessing the crisis situation and making adjustments of the communication decision as necessary. It should be noted that the rapid iteration through the steps needs to keep pace with the rapidly changing crisis environment.

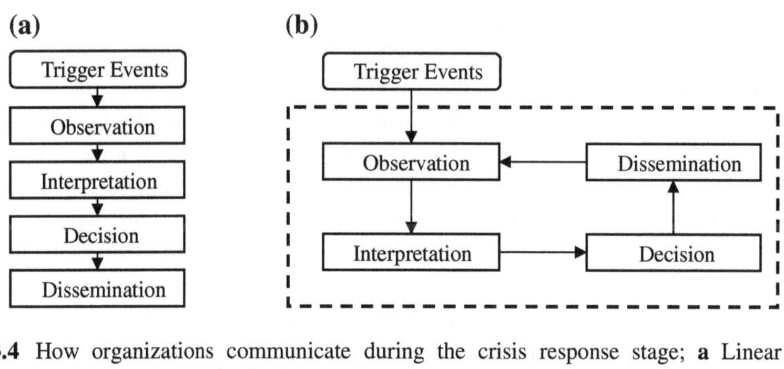

Fig. 3.4 How organizations communicate during the crisis response stage; **a** Linear crisis response communication model, **b** Cyclic crisis response communication model

3.4.2.2 Crisis Communication Within a Construction Firm

Communication may be a taken-for-granted component of organizational life, but it is by no means straightforward in complex organizations such as construction companies, which contain a variety of different types of professional, managerial, skilled and unskilled craft, and administrative employees (Dainty et al. 2006). All of these individuals need to communicate across their divisional, departmental, and professional interfaces for construction organizations. It is the classic trait of organizational structure within which employees communicate with project managers (temporary) and functional line managers (permanent) (Moore and Dainty 2001). Hence without effective communication, it would be impossible for any construction firms to compete in the challenging operating environment, especially in the event of a crisis.

Construction firms are frequently among the first respondents to major disasters and emergencies. No geographic area is immune or protected from the threat of emergencies and disasters. The major consequences and interests from crises for construction firms involve two broad categories: human and financial impact. On the human side, there is the risk of personnel injuries and/or fatalities. Employees need to be told what to do to protect themselves physically in the crisis. People are the first priority in any crisis (Coombs 2007), so instructing information need to be communicated clearly. For example, employees may need to know how to evacuate dangerous construction and find adequate shelter for cover.

It is suggested that evacuation plan and procedures should be clearly communicated to everyone on a project and reinforced by regular training and mock-drills (Loosemore 2000). The construction firms need to plan proper evacuation procedures to identify routes to safety, and reallocate dangerous machinery and equipment in the vicinity of escape routes. Also the rescue planning and procedures to retrieve injured or stranded workers from dangerous construction locations need to be developed and communicated properly. Contacts with local public safety and emergency response agencies can decrease confusion when a jobsite incident occurs (Reid 2000). A quick response can expedite medical treatment and save lives.

The importance of effective communication with all the personnel and workers at a jobsite cannot be overstated. A vitally important key to effective disaster response is a communication system that can relay accurate information quickly. Reliable communication equipments such as telephones, cellular phones, broadcast, or two-way radios need to be deployed and proper communication procedures need to be developed (CSAO 2003). Emergency phone numbers and the site location should be posted and clearly marked beside all site phones. Also it is useful to have a backup system in place, in case the communication system is rendered useless (CSAO 2003), e.g., telephone lines may be damaged.

On the financial side, the considerations for a construction company include preventing pilfering of materials and equipment, making ensure that all critical materials or supplies are not stockpiled in the same area and the major pieces of equipment are garaged in different areas. Other effects include cash flow crisis, or even bankruptcy, due to customer's inability to pay, loss of performance bonding capability, or assessed liquidated damages (Reid 2000). To be capable of surviving a crisis, a construction company needs to develop, implement, and periodically update a comprehensive emergency plan (Loosemore 2000; Reid 2000).

It is also important to maintain the IT system such as IT data, financial archives, project documentation, etc., and restore information or access backup system off site in case of any damages (CSAO 2003). Suppliers may not be able to obtain and deliver materials. The indirect costs associated with an injury are considerable. This soft-cost effect can be short-term workforce absenteeism leading to service disruptions, due to a negative swing in employee morale, fear, and a decline in productivity (Reid 2000). The factors considered by construction firms are summarized in Table 3.2.

3.4.2.3 Crisis Communication with External Stakeholders

Communication affects how various stakeholders perceive the organization in a crisis (Allen and Caillouet 1994; Benoit 1997; Coombs 2007). Mainstream communication literature advocates a holistic public relations effort which companies work constantly to maintain the corporate image in good times and bad, and continue public relation efforts to repair any damage done to their corporate image during a crisis (White and Mazur 1995). Companies need to be prepared to

Table 3.2 Concerns and interests considered by construction firms in event of emergencies and disasters

Concerns	Interests
Employee safety and health	Injuries and/or fatalities of employees
	Safety measures to protect or evacuate employees
	Reallocations of dangerous machinery and equipments to prevent potential hazards
	Loss contact with employees in the workplace
	Unsafe safety practices causing injury or fatality
Economic/cost considerations	Theft/vandalism, stealing of materials and equipments
	Damage of critical materials and/or equipments
	Additional expenses due to relocation of operations and replacement of equipments to prevent further damage
	Breach of contract, assessed liquidated damages
	Cash flow crisis or even bankruptcy, due to client's inability to pay
	Loss of performance bonding capability
Business operation continuity issues	Destruction of IT information system, failure/loss of critical computer data such as project documentation, financial archives, etc.
	Damage to utility lines
	Disruption in supply chain logistics
	Short-term workforce absenteeism

(*Source* adapted from Reid (2000))

communicate with the public immediately and address the reasonable and responsible concerns of stakeholders about the eventuality of a crisis (Millar and Heath 2004), as the timing of response to a crisis is critical. The importance of communication that adequately addresses public concern cannot be overstated, especially when the situation involves public confusion and uncertainty (Horsley and Barker 2002).

Communicating with external stakeholders or parties is vitally important for construction companies. Communication and decision making, which are made rapidly during a crisis, can have a tremendous effect on various audiences (Reid 2000). Figure 3.5 summarizes the key external stakeholders or audience in different categories of a construction company.

In the case of large-scale crises such as an earthquake disaster, efforts may require the collective cooperation and participation of multiple organizations and government agencies from local and/or state levels (Sellnow et al. 2002). Catastrophic disasters and extreme events have increased the role of the public sector in managing disasters and emergencies (Kapucu and Van Wart 2006). There are several government agencies and/or offices all over the world concerned with disaster phenomena and who provide assistance in emergency response. There include for example in the United States the Federal Emergency Management Authority (FEMA), Occupational Safety and Health Administration (OSHA), Office of Emergency Services (OES), Emergency Medical Services Authority (EMSA), Local Health Department (LHD), the Environmental Protection Agency (EPA) (Reid 2000; Shaluf 2007b).

In China, there are no national level emergency management departments like FEMA in US. Instead many departments share their responsibility for emergency management. The Ministry of Civil Affairs (MCA) generally leads natural disaster management and relief, with support from other related departments. The National Committee for disaster reduction (NCDR) is a specialized agency under the MCA, responsible for disaster mitigation and relief activities (ARDC 2009). The Office for Public Emergencies (OPE) deals with disaster emergences and coordinates central and local government. Other government agencies include the State Administration of Work Safety (SAWS), provincial or municipal government Emergency Management Office (EMO), local Public Health Department (PHD), etc.

One of the primary reasons that require a coordinated response from national and/or local government agencies in disaster situations is to provide support in a timely manner to save lives, prevent human sufferings, and mitigate severe damages (Harrald 2006). Construction managers have a key role to play because they are involved in the construction of the infrastructure and possess valuable information about their projects—the information that can be critical in disaster preparedness as well as response and recovery (Haigh et al. 2006). Hence for construction companies, it is important to make efforts to enhance communication, flexible decision making, and an expansion of coordination and goodwill with external government and emergency agencies and personnel. Timely communication contributes to informed adaptation among construction companies and government agencies and community under threats (Comfort et al. 2001).

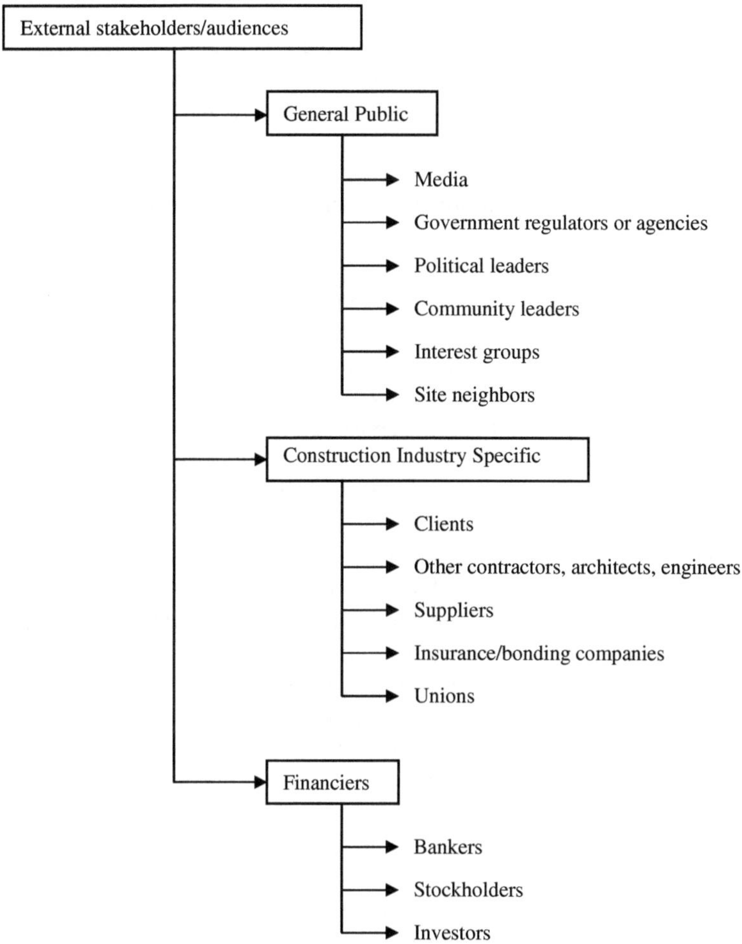

Fig. 3.5 Key external stakeholders of a construction company

However, disaster environments present a difficult context for inter-organizational and inter-jurisdictional coordination. Lack of information often leads to failure in coordination and communication (Comfort et al. 2001). Many shortcomings of the existing disaster response operation result from the fact that first responders at individual, team, and organizational levels are unable to develop a depth of understanding of the situation that would allow them to make comprehensive decisions and respond in a holistically appropriate manner (Son et al. 2007). Enhancing information search, processing, and exchange among organizations creates the possibility of increasing coordination through feedback processes. Timely communication and coordination under uncertain conditions requires an understanding of shared risk (Comfort 1999). Informed action, guided by a shared goal of protection of life and property for the community becomes a primary strategy for disaster reduction and response (Comfort et al. 2001). Further research can be done to investigate how to increase the communication and coordination effectiveness and facilitate the knowledge and information sharing among the construction companies and national agencies in the complex disaster situation.

3.4.2.4 Communication Structure and Pattern in Response to Crises

During a crisis, effective communication is essential and organizations with a track record of effective communication as an intrinsic part of their day-to-day life are most likely to turn it to advantage (Mindszenthy et al. 1988; Sikich 1993). Effective communication systems are particularly important in dealing with external stakeholders such as emergency services and the public. The media, in particular, plays an important role in constructing the public's image of events. Poor communications can therefore result in distortions of the truth, unjustified mistrust, suspicion, and irrevocable damage to customer relations.

Research has shown that it is essential to identify networks, in order to understand and manage organizational communication effectively (Carroll and Burton 2000; Emmitt and Gorse 2003). A large number of research studies on social networks explored the relationship between types of exchange connections and interdependence among actors in various network structures (Carroll and Burton 2000; Markovsky 1998). Communication structure is defined as the "differentiated elements that can be recognized in the patterned communication flows in a system" (Rogers and Kincaid 1981, p. 146). Communication can be formally or informally defined within organizational settings at the most basic level (Dainty et al. 2006). Faced with the danger of crisis, managers should understand both the formal and informal characteristics of the communication network structure and how they influence the achievement of effective crisis management.

Formal communications during a construction crisis are determined primarily by organizational structure, recognized relationship, and employment contracts that bind the project participants together. Project team members and stakeholders transmit and exchange messages across time and space in a relatively predictable pattern based on the contractual rules and standard procedures, which are designed to ensure a rapid response to the crisis by keeping the information channel clear and defining the boundaries of acceptable behaviors. In this way, the formal communication for example construction contracts has a significant effect on the behavior of stakeholders during a crisis.

At the same time, communication networks also have a significant informal element. These networks are emergent; they occur more or less spontaneously; they spring up in the day-by-day communication behaviors of individuals in an organization (Rogers and Agarwala-Rogers 1976, p. 110). In response to a construction crisis, the network of communications emerges to meet a need for significant technical and monetary change (Loosemore 1998b). Construction crises not only necessitated a change in the physical design of a building, but also required a redistribution of resources between participating organizations. In this sense, the pattern of the information exchange and communication which evolve are unpredictable as different interest stakeholders attempt to struggle for power and manipulate communication structures in ways which suit their interests.

There is considerable evidence to suggest that the ability to process information is dependent on the type of communication networks and the nature of the information processing task (Emmitt and Gorse 2003). Shaw's (1981) theoretical construction in communication networks provided the evidence to support a link between human communication structure and crisis management efficiency. The study examined the effect of communication structure on the group performance of a set of collective tasks. The experiment assembled small groups of people, requiring them to solve a simple problem that required the pooling of information. The members of each group were physically separated and only permitted to use predetermined communication channels, as illustrated in Fig. 3.6.

The results showed there are significant differences among the communication patterns in terms of their problem-solving efficiency. When the task was fairly simple, organizational efficiency developed rapidly in the centralized structure (Chain, Y, and Wheel); however, as the task complexity increased, the relative efficiency reversed completely and the decentralized structure (Circle, All-network) became the fastest. The central person in the Wheel became overloaded with information and peripheral people were less willing to merely accept the solution offered by the central person. Very similar findings on the organizational communication networks were reported by Carroll and Burton (2000) for the simple task condition experiment. However from the complex task condition, a difference in performance between those two decentralized structures was observed. The Circle structure generally did perform much better than all centralized structures; while the All-network structure took less time processing information, but had higher error rates and higher overall project costs than other centralized structures.

It was found that although the above social network studies have suggested that decentralized structures generally performed complex tasks better, these studies do not distinguished the degree of the interdependency within these communication structures. Some of the decentralized structures have a much greater degree of interdependency and connectivity than others. And to perform well as a whole, each network node or unit needs to collaborate and integrate. However, there is no

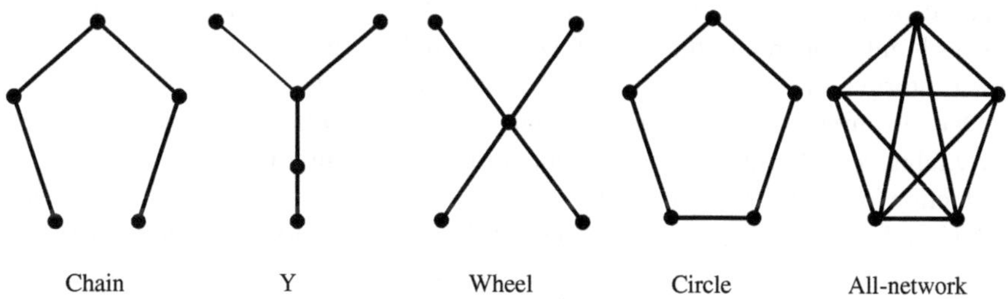

Fig. 3.6 Various structures of the communication system

guidance on how much integration needs to take place in traditional organizational theory and social network studies.

All these networks could be examined within a construction organization (Dainty et al. 2006). Therefore, it is of great importance to understand the various communication network structures during a crisis. Loosemore (1998b) suggested that the communication system structure which emerges in response to a crisis influences reaction efficiency of crisis management in construction.

3.4.2.5 Communication Challenges in Crisis Response

The crisis response stage is characterized by short decision time, stress, complexity, and uncertainty. It is at this stage that immediate and appropriate communication decisions are vital and may even affect the subsequent crisis recovery management. Crisis response communication management is rather challenging, as the information overload and channel bottlenecks (Quarantelli 1988) can cause the communication system to break down. More (1995) noted that the crisis response phase puts organization's normal communication systems and processes under enormous and additional pressure.

Crisis response strategies are a subset of crisis communication that focuses on what an organization says and does after a crisis hits. A crisis manager must review the business communication, organizational communication, and public relations literature to collect and integrate all the ideas needed to develop guidelines about what to say during a crisis (Coombs 2007).

Hale et al. (2005) reported the key communication challenges during the crisis response stage: (1) inadequate data sources at site of the crisis event; (2) inability to use routine or normal communication channels; (3) incomplete and/or conflicting interpretation of crisis data; (4) message filters based on insufficient situational expertise and organizational experience; (5) time pressure limiting exploration of alternatives; (6) conflicting resource needs of information dissemination; and (7) message format poorly designed for audience needs. In the context of construction, Loosemore (1998a) observed that at a time when effective communication was of particular importance, a crisis created conditions which made it less likely. Where there was uncertainty over responsibility patterns, information became an increasingly important source of power and was therefore more closely guarded by people. Also the crisis coping strategy by the project managers was always highly adhered to the formal, standardized procedures; the effect is to slow down information supply and cause a significant amount of frustration.

One example is the Piper Alpha disaster (1988, available at http://www.fabig.com/Accidents/Piper+Alpha.htm) that occurred on an offshore oil production platform resulting from large-scale explosion and fire in July 1988 in UK and killed 167 persons and cost billions of dollars in property damage. The report of the public inquiry into this accident exposed weakness of the crisis communication management, and in particular the unreliable communication between the production shift employees. The production shifts did not follow company procedures on the Permit to Work system and information about the current maintenance status was not sufficiently communicated and passed from one shift to another (Fraser 2009). This crisis communication failure highlights the limitation of the conventional linear communication model that information, communication symbols, and rules are not passively transmitted as intended between the communication parties. An appreciation of the complex interplay of meanings and symbols of human communication and shared understanding between both parties are vitally important for improved performance. Another cause in part to the communication failure is the disruption of the communication channel. During the accident the emergency communication central radio room was damaged, and not able to be used to communicate to personnel for evacuation. Employees on board were given no further evacuation instruction but waited in smoke-filled room. This also indicates the inflexibility of the decision making in response to the unexpected situations.

Another example is the Singapore Nicoll Highway collapse accident in April 2004. From the report by the Committee of Inquiry (COI) into this accident (available at http://www.mom.gov.sg/publish/momportal/en/press_room/press_releases/2005/20050513CommitteeofInquiryconcludesstringofcriticaldesign errorscausedcollapseatNicollHighway.html), the deficiencies in the project management includes problems in the intra- and inter-chain of command and communication both within the contracting organization and between contracting organization and other parties. It is recommended to establish a proper chain of command and reporting system to facilitate the proper flow of information on site, and maintain effective communication between top management and workers/staff. This should also be part of the culture in the workplace. It is also criticized in this accident that heavy reliance on past experience was misplaced and not properly adapted to other localized incidences in the project. So standard but undifferentiated measures were ineffectual, linear causality assumption cannot be always valid in the real world. In addition, the staff/works should be empowered to make effective decisions on unsafe workplace practices, as well as remove or eliminate work hazards.

Hence, the purpose of the crisis response communication management needs to be redefined by complementing this with a broader concept of the organization's adaptability and resilience in the complex environment. The communication system should enable the organization to become resistant to perturbations and enhance its capacity to restore itself after a crisis (Paraskevas 2006); in the sense to fulfill the business continuity challenges (Low et al. 2008).

3.5 Summary

The chapter summarized the mainstream research which has been carried out on the crisis management and communication management so far. In addition, this chapter also examined the communication challenges and issues during the crisis response stage, which is characterized by short decision time, stress, complexity, and uncertainty, where an immediate and considerate response and communication decision making is vital. It was noted that the conventional linear communication models failed to account for the interactive and self-organizing aspects of the communication process in the situation of a crisis. Chapter 4 proposes a new conceptual framework for crisis response communication management to fill this lacuna.

Chapter 4
Conceptual Framework

4.1 Introduction

Complexity theory provides an understanding of the social behaviors of project teams and the networks of people involved in and around a project. The consequence of accepting complexity theory is to shift the focus of project management from the object of the project to the actors and their interrelationships and interaction involved in the project. Drawing upon relevant concepts in complexity theory and organizational studies and management, a conceptual framework for understanding the underlying pattern of communication behavior and decisions of human systems in response to a crisis is proposed in this chapter.

4.2 Complexity Theory and Crisis Communication

4.2.1 A Complex Conception of Communication

Complexity theory excels in explaining how complex behaviors of adaptive systems emerge from simple rules of interactions among members of the system. Complexity involves unpredictable behavior, interaction and feedback loops, decentralized decision making, and interdependence of various simple systems that, together, make a complex system. Social scientists have experimented with applications of complexity theory in understanding human social systems (Kaufman 2006). Human systems are intrinsically complex in the sense that they are able to self-organize, to influence each other, and be influenced in turn, and this reciprocal influence can change ideas, behavior, ways of thinking, working, and relating—that is, human is able to co-evolve, to self-organize, and to creating something new that is emergent in the sense that it could not have been predicted at the outset (Mitleton-Kelly 2005). The interaction of all the subsystems of a complex system and role of the relationships formed, as well as the creation of

Z. Ying and L. Sui Pheng, *Project Communication Management in Complex Environments*,
DOI: 10.1007/978-981-4560-64-1_4, © Springer Science+Business Media Singapore 2014

information and knowledge through these interactions, form the basis of the complexity approach (Cilliers 1998, p. 10).

Stacey (2003) declared that communication is a complex process of relating—a chain of patterned responses that provide the context for an individual action across space and over time. From the perspective of complex responsive process of relating (CRPR), organization is an emergent property of individual human beings interacting together centered on the use of language simultaneously for conversation and to negotiate social status and power relationships. Communication by means of evolved language is a defining characteristic of human beings, distinguishing them from other species of animal (Kauffman 1993). Central to the theory is the recognition that communication is a complex process involving both the words that are spoken and the response that they elicit. Weick (1983) opined that humans seek to reduce equivocality and uncertainty by organizing. Communication is an act of organizing; it is human nature to create order and organization symbolically by coming in contact with the material world. Organizations, language systems, conceptual systems, indeed, anything that brings order come as a result of the interaction between the existing symbolic system and a real material world. The result is a combination of symbolic and material order and disorder which have self-organizing characteristics.

The conventional communication theories reduce communication to a process or an act that occurs within a system. The system acts as a transmission mechanism. The system exists without communication, though it might not be considered dynamic. When it is dynamic, communication is the movement of useful and relevant information. Noise is information that the system cannot use. There is a message in the medium as well (Shannon and Weaver 1949). The complexity of communication says that information is the system. Complexity theory contends that without information living structures do not exist. Communication is the act of the life. From this perspective, the act of communication cannot be separated from the system. The information is life-giving, and the movement of information or communication is essentially life itself. Information not only informs the others, but it creates itself (Botkin 1990; Kauffman 1992). Understanding information as a dynamic process has important consequence for the way we deal with it in our social lives, especially in our business organizations. Complexity theory holds that information does not exist independently of a system, it is not something out there that the system picks up from the environment. Instead, the system selects which disturbances to notice and consequently, creates information and assigns meaning to it through structural coupling. Thus communication is not a process by which organizations exchange information, but instead, it is the organizational system itself used for coordinating behaviors.

Complexity theory argues for a more holistic view and practice that one-way relationships are not present in natural systems, nor are they appropriate in the conceptual system (Hayles 1990). In societal terms, this would apply to people in any context and would imply that a person or group of people derive their meaning from the relationships they have with other individuals or groups in their environment. Communication in the complexity paradigm does not happen within a system; rather it is the system. The act of communication defines and describes emergent systems. Thus, communication is the dynamic emergent act of creating new information in the overlapping spaces of two or more dynamic systems, or subsystems. The complexity paradigm views communication not as linear information exchange or downloading. Every time a text interacts with a reader, a manager with an organization, or an audience with a culture, new information is created. The creation of new information involves a pattern that is not predictable because it includes randomness generated by human agency in all human communication systems and media.

Traditional views of communication suggest that randomness and chaos result from the lack of communication or information. A complexity definition of communication says otherwise. When randomness is understood as maximum information, it is possible to envision chaos as the source of "all that is new in the world" (Hayles 1990, p. 51). Thus a complexity view of communication argues the chaos, the lack of knowing, the disorder, and the ignorance, all have a place in creating a context for order and perception. In other words, ignorance can be considered as potentially essential state before an act of knowing and communication can occur.

Traditionally the interpretation of data and information was done by management, which in turn led to filtering, subjectivity, exclusivity, and over-control (Wheatley 1994). Wheatley (1994) suggested that there is interdependence between different subsystems in an organization, which indicates that all the subsystems should take part in the processes of the system. Participation could add to the richness of information, shared responsibility, more trust and transparency and, ultimately, to more healthy relationships. This interdependency and participation in turn imply relationships, the sharing in decision making, as well as in the dissemination and interpretation of information throughout the organization. Hence, the traditional linear theories of communication, persuasion, and management do not really explain the dynamic and even chaotic communication environment in most organizations today.

In general, the complex interaction patterns constitutive of human societal systems and organizational systems are centrally expressed in communications. Organizations and individuals create complexity when they use syntax (rules, codes or symbols) to translate semantics (meaning). Communications underlie the political, economic, and social process governing patterns of interaction and the resultant social aggregation process (Kaufman 2006). Therefore, it can be extrapolated from one of the principal focuses of complexity theory and study which is common to all living systems, called complex adaptive systems, to understand and analyze the crisis response communication systems in the context of construction organizations.

4.2.2 Crisis as a Bifurcation Point of the Organizational Communication System

Bifurcations refer to a system's condition or behaviors suddenly dividing or branching into two different or merging part behaviors (Aula 1996). Crisis creates the opportunity to redesign, revise, or rebuild the human environment damaged by the event. The systematic changes that emerge in response to a chaotic environment are referred to as bifurcation. All complex systems, even those with the appearance of stability and order, have the potential for bifurcation. Those systems with high states of exchanges and connections with their environments may be typified by higher levels of instability and periodic bifurcation. Crisis events and behaviors, also associated with environmental dependence and change, are often described in chaos theory as points of system bifurcation (Seeger 2002; Murphy 1996; Matthews et al. 1999).

A crisis can be considered as a bifurcation point in the organization's history which irreversibly changes its culture and business (Murphy 1996). These points are not random but occur because of accumulated flaws or problems within the system. Effective communication management is a critical tool in the management of a crisis situation (Marlow and Wilson 1997). Organizations often experience information meltdown during a crisis, or management could decide to take total control of all information and only feed through what they consider important, necessary or safe. It is at this point that open, free, and total flow of information is crucial (Flower 1993).

Communication processes are viewed as critical factors in the stable, routine operation of organizations as well as in unstable chaotic points of a crisis. The underlying patterns and processes of communication may be factors that bring about system stability, order, and balance even in the face of chaos. Crisis stakeholders may communicate in new ways, exhibiting high levels of cooperation, creative problem-solving, and collaborative decision-making. Hence, crisis can also be viewed as a bifurcation point of the complex organizational communication system, where disequilibrium in established orderly communication operations and procedures occurs.

Crisis response communication can serve to activate or accelerate the functioning of various attractors following a crisis, thus helping to constitute system reorganization and renewal. The next section discusses that the crisis response communication can be viewed as a complex adaptive system (CAS), which is able to change its internal communication structure and pattern in order to learn and adapt to the dynamic and changing situation.

4.2.3 Crisis Response Communication Systems as CAS

The term CAS refers to the systems that are capable of changing themselves in response to a changing environment, and that their order emerges from lower level interactions (Anderson 1999). From a complex systems perspective, it is the interaction of the components of the system, rather than the system itself, that is a source of influence and which learns to alter its functioning over time. Learning behavior occurs through system adaptation and change and the selection of competing schemata, which is the result of direct cybernetic adjustment, schema alteration (of individual agents or the entire system), or the elimination of the less fit (Gell-Mann 1994). The complex behavior of systems arises from the interrelationship, interaction, and interconnectivity of elements within a system and between a system and its environment (Mitleton-Kelly 2005).

Cilliers (1998) explained that a complex system is not constituted merely by the sum of its components, but also by the intricate relationships among these components. It is not merely the way that can be described as simple or complex regarding a system, but complexity results because of the interactions and relationships between subsystems within the system. In terms of communication system, the communication behavior cannot be understood if this is not placed in relation to other behavior and in an organizational or social setting. A word cannot be studied without seeing its relational language as a whole and to other words. An individual communicative behavior cannot be understood without understanding its relationship with other individuals within the system.

This CAS definition can be applied to the crisis communication systems for construction project organizations, where systems of objects, i.e., project members, teams, organizations, industry agencies, communicate and interact with each other in a nonlinear way, based on various types of contractual relationships and professional rules, exchange information to enable the organizations to operate more effectively and adaptively in the face of crisis. This process of interaction through communication in the event of crisis indeed describes the fundamental characteristics of the behavior of complex adaptive systems.

Crisis response can be viewed as a complex system with fuzzy boundaries and diverse agents who come from several different parts of the organization and serve one or more crisis response tasks: signal detection, prevention, damage limitation, recovery, learning, and redesign (Mitroff 2005). The intricate interactions among various agents within the organization and between stakeholder organizations are created and sustained through the communicative process. The interactions are "local" (Stacey 2001) in the sense that they have a relatively short range, primary affecting neighboring agents (Gilpin and Murphy 2008). No individual agent has complete knowledge about the behavior of the system as a whole, only the information received locally. Cilliers (1998) also noted that the influence of the local interaction in the individual base gets modulated along the route and will be enhanced, suppressed, or altered in a number of ways. Gilpin and Murphy (2008) suggested that project managers can apply this principle in the crisis management and that they should rarely focus on a single individual in the environment but look for patterns built up from individual actions, such as customer response, or media coverage.

In a complex system, the interactions are iterative and recursive (Stacey 2001). The effects of the interaction among agents are reciprocal or looped, in the sense that they can feed back at any point in the system, positively or negatively. Crisis usually generates a greater degree of pressure and urgency to resolve a problem (Loosemore 1998b). Positive feedback had characterized a majority of supportive, problem-orientated communication encouraging change, while negative feedback had a majority of obstructive, defensive communication encouraging stability. In terms of crisis communication management, project managers should demonstrate a constant sensitivity to the types of feedback. The feedback mechanisms require that the components of complex adaptive systems learn and use their learning as the basis for action and communication with each other. Learning consists of a system's ability to scan its environment to detect variations and apply a set of rules in order to adjust its behavior appropriately. It should not be just understood as acquiring new information.

The communication interaction of the individuals within a complex system is also self-referential (Stacey 2001). The system's adaptive learning ability is always consistent with the history of its past structural change and evolving experience. When the crisis happened, people's preconceived beliefs about other's interests and motives can have significant influence on their behavioral response and communication. For instance, the poor communication and interaction between the architect and the contractor had been transferred from a previously unsuccessful project relationship where the contractor had gained a reputation for being claims-conscious. That is because the architect had developed a stereotyped view of the contractor's motives and behaviors, leading him to initiate an inflexible and self-protective communication strategy very early in the crisis response stage.

The results of the interactions are unpredictable. Probably the best known and most fundamental concept is the nonlinearity of complex systems, built on the principle of "sensitive dependence on initial conditions", also known as the "Butterfly effect" (Lorenz 1963). It describes the situation where minute changes in the starting condition can have major and unpredictable consequences in nonlinear complex systems (Holland 1995; Murphy 1996). In the midst of a crisis, small variance in the message form and content, message interpretation, message distribution, or information processing may produce unpredicted huge fluctuations, leading the whole communication systems into a chaotic state. This nonlinearity principle also suggests that precise and accurate communication about human behavior and decision making during the crisis response stage is impossible. Managerial intervention was often used to bring about equilibrium, but communication structures which were forcibly imposed to suppress the interests of other project members were found to be dysfunctional (Loosemore 1998a). Hence the traditional cause-and-effect linear communication model in an attempt to predict and control is seriously questioned.

Construction crises not only necessitated a change in the physical design of a building, but also required a redistribution of resources between participating members and organizations (Loosemore 1998a). Information sharing and exchange in a communicative process plays a critical role in facilitating these changes and reallocations toward a more balanced and win–win mode. In practice, the pattern of exchange resulting from "power struggles between different interest groups who attempted to manipulated communication structures in ways which suited their interests" (Loosemore 1998c, p. 29) was highly unpredictable and continuing to change. This unpredictability characterizes the complexity of the communication pattern of construction crisis.

Considering the crisis response communication as a CAS challenges the traditional definition of plan and control in the context of crisis response management. Conventional linear theories of communication, persuasion, and management imply authority and central command. Conversely in the complex systems, control is distributed among the system's components or agents in the way that each project participant in the construction project organization is actively involved in making response decisions locally with his/her own cognitive ability, using multiple feedback loops. Indeed, human systems are complex in the sense that they are able to self-organize, to influence each other, and create intricate networks of relationships sustained through communication.

4.2.4 *Self-Organizing and Adaptation in Crisis Response Communication*

From a self-organizing system perspective, the structure of the system is emergent, which is a result of the interaction of autonomous individual agents and subsystems. Structure emerges from randomness; system produces order from the edge of chaos (Holland 1995; Waldrop 1992). The system must operate in a far-from-equilibrium condition, analogous to the edge of chaos, since nothing novel can emerge from systems with high degrees of stability (Horgan 1996), with dissipative structure (Prigogine 1984) that dissipate energy while recreating themselves into new forms of organizations. In the context of construction projects, self-organization may occur all along the project life cycle, for instance when a project organization's existing formal communication system is not effective when a construction crisis interrupts the flow of information and disrupts the normal operations. From another perspective, crisis can also be viewed as a bifurcation point of the complex organizational communication system, where disequilibrium in established orderly communication operations and procedures occurs.

Self-organizing processes can be expected to arise following bifurcation. This self-organization is manifested and influenced through strange attractors, and eventually reconstitutes the system at a higher level of order and complexity. Upon the bifurcation points, the complex system may either break down leading to the complete chaos or break through to one of several new states of order and higher levels of complexity, which will emerge from self-organizations, arising from the spontaneous reallocation of energy and action (Kauffman 1993). This self-organization is a force of antichaos (Kauffman 1996) using the feedback loops mechanism within the systems. Importantly, how the system will behave cannot be determined by some central controlling mechanism, but each agent or component's perception of the situation. Comfort et al. (2001) described a case of spontaneous self-organization following a natural disaster. After a 1987 southern California earthquake, traffic lights stopped functioning. In a spontaneous re-emergence of order, individual drivers exited their cars and began directing traffic. Even the extreme chaos and uncertainty of the 9/11 disaster created spontaneous self-organization at a variety of levels (Seeger 2002).

In terms of crisis communication system, the collapse during a crisis generates new communicative structures and relationships, understandings and procedures in a manner consistent with self-organization and emergence characteristics. In terms of crisis response communication, complexity theory's emphasis on the patterns of self-organization also suggest that attention should move beyond immediate and short-term dynamics of crisis, to larger patterns of self-organization, reconstitution, and renewal in a long-term frame. The role of crisis communication should be examined more broadly in growth and renewal of an organization following bifurcation and the organization's business continuity management plans (Low et al. 2008).

Morgan (1997) described the process of reorganizing that analyzes the environment and the patterns of relationship and communication network between actors and explains organizational adaptation and development. The process of reorganizing may not be driven from the top management and authority such as the project manager, but may occur as team members or project participants interact with one another based on the needs of the moment. The project participants create an informal communication network that is superimposed on the existing formal organizational structure. This informal communication network in response to a crisis hence emerges as the new communication structures are constructed between participants based on the information sharing and demands. Indeed, because of the nonlinearity and unpredictability of the CAS as discussed in the previous section, combined with the ambiguity of meaning in all symbols, changes, and commands driven by the top management are usually unlikely to produce desired effects without local individual's coordination.

With the self-organizing and adaptation property of the complex systems, they continually exchange energy with their environment, allow spontaneous behavior and create new patterns. This allows them to poise in a state that is far-from-equilibrium, "on the edge of chaos" (Langton et al. 1992; Kauffman 1993; Stacey 1996), where order and disorder, stability and chaos co-exist. It is at the edge of chaos that systems have the greatest potential for change, growth, development, and creativity (Stacey 1996) and have the capability of complex learning or double-loop learning. Hence, the next section highlights the insights of complexity theory to provide a conceptual framework to analyze and discuss these issues in the context of crisis response communication management.

4.3 A Complexity-Informed Model for Managing Crisis Response Communication

As mentioned in Chap. 3, crisis studies have claimed increasing attention to the interaction between organizational members and stakeholders and the development of strong, positive relationships. They rely on the concept of relationships as repeated local interactions that allow the organizational members and stakeholders to adapt to each other in response to the crisis. This process of interaction through communication in the event of crisis indeed describes the fundamental characteristics of the behavior of complex adaptive systems. In applying complexity theory to communication management in the context of crisis response, the interactive patterns between individuals and organizations underlying the shifting social aggregations can be better understood (Kaufman 2006).

As the lessons from complexity theory show that creativity and double-loop learning cannot occur within a system's state of equilibrium; a CAS self-organizes at the edge of chaos. Therefore, when confronted with the discontinuities or sudden unexpected changes in the environment such as a natural disaster or other kinds of crisis, the communication system and the organizations themselves must be able to re-emerge out of the chaotic state, and eventually reconstitute the system to adapt to higher performance levels on the system's current or emergent fitness landscape (peak level) (Paraskevas 2006). In this sense, the communication system has good adaptability to the changing environment, capable of migrating to the edge of chaos. A fitness landscape is illustrated in Fig. 4.1 (adapted from Stacey (1996)). The fitness landscape covers the array of all possible survival strategies open to a complex subsystem. The shape of that landscape is defined by survival strategies that all other subsystems within the environment are pursuing, as all the subsystems interact to coexist. To survive and be more flexible and responsive to the complex dynamic environment, a complex subsystem must be able to evolve to discover the highest peak, a journey across the fitness landscape (Stacey 1996).

In this section, a complexity-informed model is proposed to analyze and explain the communication system in the situation of a crisis. The conditions that enable the crisis response communication system to be capable of poising at the edge of chaos and the control parameters (CP) determining whether the system is at the edge of chaos would be explored.

4.3.1 Control Parameters

Stacey (1996, p. 99) pointed out that there are critical points in parameter values, at which a system occupies the space for endless variety, novelty, and creativity with enough disorder to prevent the system from becoming trapped in some local equilibrium to the detriment of its long-term development to higher fitness peaks,

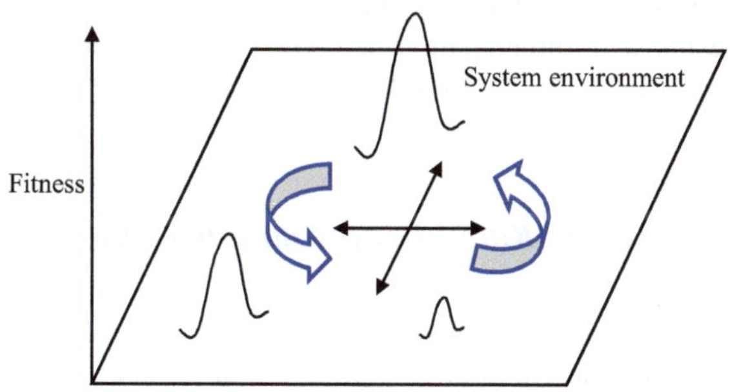

Fig. 4.1 Example of fitness landscape

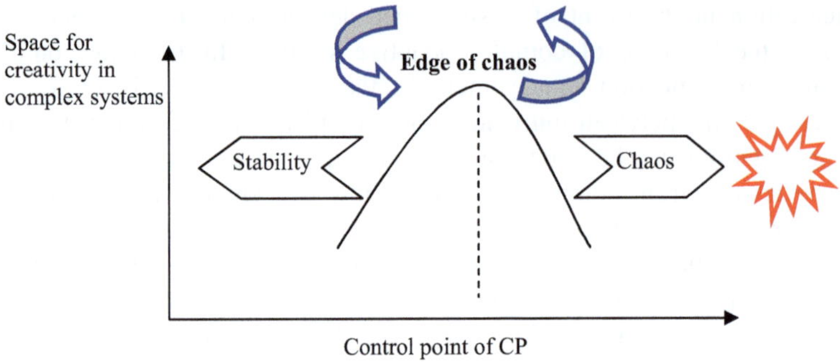

Fig. 4.2 Illustration of CP' influence on complex system performance

but also with sufficient containing structure and order to prevent it from falling apart into patternless behavior. As these CP are turned up, the adaptive nonlinear feedback networks pass from stability through a phase transition at the edge of chaos into complete chaos [See Fig. 4.2 adapted from Stacey (1996)]. Five CP influence and indicate a CAS's ability to migrate to the edge of chaos: (1) the rate of information flow; (2) the richness of connectivity between agents; (3) the degree of diversity among the agents of the system; (4) the level of contained anxiety; and (5) the degree of power differentials. The latter two CP are applicable to human systems. These CP are borrowed and applied here to the analysis of the crisis response communication process.

As mentioned previously, the organization can be viewed as the organizing process achieved through communication. Drawing upon knowledge in organizational studies and management, including a number of interconnected domains of strategic organizational design, social network, learning organization theory, and knowledge management (Burton and Obel 2004; Miles and Snow 1978; Stacey 2001; Wasserman 1994), specific organizational variables (OV) that influence the level of CP were identified. These OV are grouped into six major categories: (1) organizational structure, (2) organizational cultures, (3) the information technology capability and information management system, (4) style of management and leadership, (5) sense-making capability of organizations and members, and (6) skill capability of members of the organization.

The complexity-informed conceptual framework is proposed as seen in Fig. 4.3. The next section provides analysis of the impact of the CP on the crisis response communication performance, and then discussed and evaluates how the identified OV affect the CP within the context of crisis response.

4.3.2 CP1: Rate of Information Flow

The flow of information in a system is what keeps a system alive; it also builds strength into a system (Youngblood 1997). Stacey (2001) treated the human

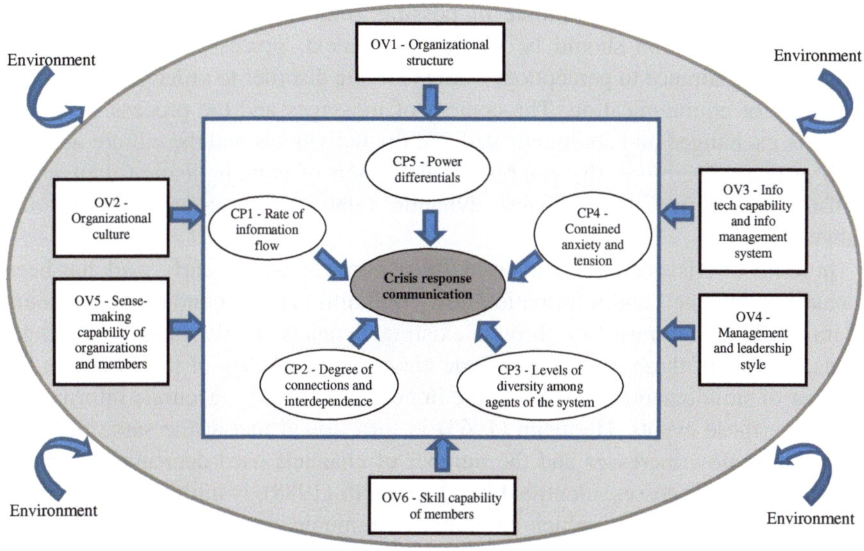

Fig. 4.3 Conceptual framework for managing crisis response communication

analogs of information flow in complex adaptive systems as the number of connections between themes that organize the experience of relating. In other words, the richness of the themes organizing the experience of relating has an impact on the dynamics of conversation. When relationships between people are organized by a small number of loosely connected themes, conversational patterns become repetitive and predictable, and they lack any spontaneous and free-flowing qualities. On the contrary, when the conversation is organized by many themes triggering many associations with others, communication is likely to become highly disorganized (Stacey 2001). Accordingly, when the project participants have overflow information exchange among them in a short period, too much "noise" may be created which may be too overwhelming for them to absorb, resulting in completely chaotic situations. Conversely when project participants do not have much cross-disciplinary information at hand, they may not be able to mutually communicate with each other in an appropriate way.

The traditional view of management in terms of communication is that information is power and it has to be controlled and "fed to employee in little doses" (Flower 1993). The interpretation of data and information was done by management, which in turn led to filtering, subjectivity, exclusivity, and over-control (Wheatley 1994). This worldview implies that perceptions must be managed by feeding the "right" information and withholding information that might lead to disorder and chaos (Youngblood 1997). The immediate reaction to disorder was to control the information. Hayles (1990) noted that according to complexity theory, randomness in relationships produces information and the need for communication. "Information cannot be calculated for a message in isolation. It has meaning only with respect to an ensemble of possible messages (Hayles 1990, p. 52)". Thus, communication should be studied as context, process, and content. The move from ignorance to perception, from symbolic disorder to order, is essential in the study of communication. The content of messages and the process by which they are exchanged and communicated and the individuals and the culture are also both critical. Therefore, the complexity definition of communication demands a holistic perspective that explains dynamic relationships among systems and subsystems.

In terms of crisis communication in the response stage, the early work has been focused on channels and information flow. Williams (1957) found that the amount of information that must flow through existing channels greatly increased during a crisis. Causes of these increases include elevated complexity of the situation; the number of simultaneous events; and the importance of quick, accurate information regarding these events. Hermann (1963) in turn, found that at the same time the information flow increases and the number of channels used decreased. The end result of these processes, identified by Quarantelli (1988), is information overload and channel bottlenecks, which can cause communication system failure and/or the loss or delay of relevant information reaching the appropriate participants. By sharing information and disseminating it quickly and accurately within the organizations and between organizations and their key stakeholders, crisis management will have greater success.

Organizations need information flows to function and strive to create information flows to be effective. Organizational theory holds that organizational structure ought to match the demands of the task and environment (Burton and Obel 1998). Centralized structure allows rapid information processing with limited communication demands (Galbraith 1974). In other words, decentralized structure, however, requires more communication. When the task is simple and the project environment is relatively stable, centralized project structure will be more efficient than decentralized structure. For organizations in complex environments performing complex tasks, this means relying on a sufficient level of internal complexity based on differentiation and specialization (Lawrence and Lorsch 1967). Increased differentiation requires increased integration to coordinate the disparate functions, which increases the information processing load and the rate of information flow. For this reason, decentralized structure is expected to perform better than centralized structure. Galbraith (1974) offers various strategies that either (a) increase information processing by becoming faster within the hierarchy or bypass the hierarchy with liaison and integrative mechanism, or (b) decrease the need for information processing by rearranging the tasks into more nearly independent units or standardizes the operation.

When the organization is threatened by environment changes such as a natural disaster crisis as well as internal changes such as organizational transformation and structural change, the need for communication increases (Gruing 1992). Organizations will redesign themselves to become more open and make their structures more horizontal in order to adjust to fast-changing environments in the Communication Age (Gouilart and Kelly 1995). During a crisis in which information flows far higher than can be expected, decentralized organizational structure is expected to perform better than the centralized, hierarchical structure (Loosemore 1998b). In another terms, communication and control within hierarchical structure should flow more horizontally than vertically in the complex system environments (Sherman and Schultz 1998). Increased flows of information will occur if management is able to listen without judging, allowing authority, and responsibility for that authority to rest where it belongs (Kelly and Allison 1999). The more rigid and bureaucratic the project organizational structures and communication processes are, the less connectivity and interactivity may occur between team participants. As a result, project members follow the predefined procedure to implement and transfer the information about their crisis activities, the information flow are more one-way and the potential errors and areas for reinforcement may not be recovered in a timely fashion.

Project managers not only try to assemble an appropriate organizational structure and management procedures, but also try to employ information technology which may include various types of communication devices such as computers and software, networks, Internet and other various measurement technologies (Hapgood 2004; Ravichandran 2005). New information technology changes the way people can communicate. The development of information technology has greatly increased the connectivity level within the organization and among organization and various stakeholders. Information technology facilitates both connections to disciplinary information and explicit knowledge that resides in each project participant and the real-time exchange of information. Consequently it helps to shorten the feedback loop between project members. For example, an Intranet communication system with e-mail capability helps information sharing about the crisis events in place. A dedicated crisis website can serve as an information hub for the project members to monitor the overall situation. In addition, Web-based collaboration tools also facilitate both synchronization and asynchronization of collaboration between different professionals. To improve the connectivity of the project organization, a project team must have the capability to effectively manage and communicate the project information.

A proper information management system is important to ensure that enough relevant information is available at the right time to make the right decision (Burton and Obel 2004). There are numerous means to exchange information, including the development of rules and procedures and information exchange in the form of regular meetings, integrated systems, or other liaison activities. Further the media richness of the information system is of interest (Burton and Obel 2004). Media richness relates to the type of information that can be processed and the type and speed of feedback. For example, face-to-face meetings have higher richness than short writing notes. In the complex condition of crisis, while the actual decision to mobilize emergency operations remains the function and responsibility of the project manager or crisis manger, the appropriate use of information technology can significantly improve the validity, timeliness, and also the accuracy of information available for them to manage such events. Eventually, increased effectiveness in organizational performance in the complex environment is expected to be achieved. Segel (2000) suggested that since complex systems influence not only their own state but also the state of their external environment, an informational network of sensors that sense these influence is needed. Lack of available and reliable data is one pragmatic reason that organizations do not learn from their crisis experiences (Elliott et al. 2000).

Communication is the dynamic emergent act of creating new information in the overlapping spaces of two or more dynamic systems and subsystems. Whether a project organization is able to move to the edge of chaos or not depends on its ability to operate as a dynamic communication network that thrives on information and knowledge sharing (Lesser and Prusak 2004). Construction organizations are socially complex arrangements consisting of relationships and interactions involving individuals and groups with diverse mindsets and interests. Knowledge flows among social actors are heavily dependent on both social structure and the motivation to engage in knowledge sharing (Swan et al. 1999; Hislop et al. 2000). A knowledge management and knowledge sharing method must, therefore, hold at its core the notion of the interaction of individuals (Fernie et al. 2003).

An organization's ability to develop its culture and behavior around the sharing of knowledge is called the organization's information ecology (Schage 1997). Marra (2004) used two crisis cases, one at AT&T and one at the University of Maryland, to show that successful crisis communication depends on an organizational culture of open communication. Marra's (2004) study shows that rules for interaction are developed in the micro-level with interaction between agents as a situation unfolds but are patterned by macro-level organizational expectation and context, "a supportive organizational philosophy" (Marra 2004, p. 324). Hence organizational culture fostering the interaction and interaction between project participants is another critical element within which the project-related messages in response to crisis are communicated for effective project communication. A high level of trust and good interpersonal relationship also facilitates the interaction and communication between project participants.

In order to achieve a more decentralized information flow, project managers should create and maintain channels for discourse and they should not manage information but rather open all information systems to allow self-organizing of communication. Project managers should try to understand the complexity of the environment and the interaction of all the different components of the organization, rather than just measure it. Lewin and Regine (2001) suggested that a true leader allows new processes to emerge freely within certain guidelines. In order to achieve an adaptive system, a style of leadership which encourages the communication and interaction between project participates is required. Project participants should also have the ability to effectively convey their opinions and manage project knowledge and information.

In fact, leadership role typology is related to the information network, information processing, and decision making (Mintzberg 1980; Luthans et al. 1988; Yukl 1981). Burton and Obel (2004) operationalized the leadership and management style into six dimensions: (1) preference for delegation; (2) level of details in decision making; (3) reactive or proactive decision making; (4) decision-making time horizon; (5) risk reference; and (6) motivation and control. Miller and Toulouse (1986) suggested that the leadership style has more effect on the strategy and structure when the organization faces a dynamic, unpredictable, and changing environment than when it operates in a stable and simple environment. Their empirical study of 97 Canadian firms generally supported their hypotheses.

Leavitt (1951) and Shaw (1954) suggested that crisis management efficiency would be influenced by management's ability to control the structure of people's communication patterns. They found that the influence of organizational centrality depended on the nature of the problems an organization faced, more complex, nonroutine problems such as crises demanded less centralized structures to alleviate the potential for information overload (Loosemore 2000). Thus during a crisis, the crisis manager must seek to control the degree of the centrality to which the project organization is centralized around particular individuals, or stakeholders. Mintzberg (1976) also found that during a crisis, people tend to tighten control, the consequence of which is dysfunctional behavior.

Construction contracts are designed to ensure a rapid response to crises by defining the boundaries of acceptable behavior and keep information channel free from the irrelevant information. However, Sagan (1991) argued that the influence of contractual rules and procedures diminishes during a crisis. Thus, the network of communications that emerges in response to a construction crisis has a significant informal element that could be dangerous if managers relied too much on contracts as a means of control (Loosemore 2000). Stacey (1996) explained that when it is impossible for formal systems (e.g., contracts, standard rules) to retain enough information, then it become necessary to use the shadow informal system which can retain faster flows of information because of its informality and when the information is retained and acted upon at local levels. Hence construction organizations may create ad hoc, informal networks of communication in response to the crisis, through which information may flow more easily.

The above discussion is summarized in Table 4.1.

4.3.3 CP2: Degrees of Connections and Interdependences

The dynamics of adaptive nonlinear feedback networks are also sensitive to the degree of connectivity between agents in a system (Stacey 1996). Connectivity and interdependence is one aspect of how complex behavior arises. In a human system, connectivity and interdependence mean that a decision or action by any individuals (group, organization, institution, or human system) may affect related individuals and systems (Mitleton-Kelly 2005). When this influence is in one direction, adaptation of one entity as a response to the influence of other entities may be seen; when the influence and response are reciprocal, co-evolution or change in all interacting entities may be seen (Mitleton-Kelly 2005). While the connectivity is a critical variable that influence a network's ability to self-organize, the degree of the connection between agents and the closeness of relationships

Table 4.1 Measurable OV affecting CP1

Organization variables	Operationalization
Organizational structure (OV1) (Gouilart and Kelly 1995; Kelly and Allison 1999; Loosemore 1998b)	Flat hierarchy structure (Galbraith 1974; Loosemore 1998b; Sherman and Schultz 1998)
	Highly bureaucracy structure (Kelly and Allison 1999)
	Ad hoc decentralized structure (Gouilart and Kelly 1995; Loosemore 1998b)
	Routine administrative structure (Burton and Obel 2004)
	Allows for integration of downstream activities (Galbraith 1974)
Organizational culture (OV2) (Lesser and Prusak 2004; Marra 2004)	There is learning climate encouraging knowledge and information sharing (Lesser and Prusak 2004; Swan et al. 1999; Hislop et al. 2000)
	There is a collaborative culture fostering interaction and relationship between project participants (Fernie et al. 2003; Marra 2004)
Information technologies capability and information management system (OV3) (Elliott et al. 2000; Fernie et al. 2003; Segel 2000)	Web-based technology is used to support project communication (Hapgood 2004; Ravichandran 2005)
	Project participants have enough relevant information they need to do their job at the right time (Elliott et al. 2000)
	Project members meet face-to-face or via teleconference on a regular basis (Burton and Obel 2004)
	Feedback and response to the issues are prompt in a timely fashion (Burton and Obel 2004)
	Innovative integration information system is adopted (Hapgood 2004; Ravichandran 2005)
	The same information is available to the appropriate parties (Burton and Obel 2004)
Management and leadership style (OV4) (Leavitt 1951; Lewin and Regine 2001; Miller and Toulouse 1986; Shaw 1954)	A flexible communication management process and procedures to allow for open communication (Kelly and Allison 1999)
	A rigid and centralized communication management process and procedures (Kelly and Allison 1999)
	Contractual rules and procedures are strictly followed during the response stage (Loosemore 2000; Sagan 1991)
	Project manager or leaders are involved in great details in decision making (Burton and Obel 2004)
	A decentralized decision-making process (Burton and Obel 2004)
	A proactive decision-making process (Burton and Obel 2004)
	A long-term decision-making vision (Burton and Obel 2004)
	Preference of delegation of authority leadership style (Burton and Obel 2004; Mintzberg 1976)
	Preference of control of authority leadership style (Burton and Obel 2004; Mintzberg 1976)

(continued)

Table 4.1 (continued)

Organization variables	Operationalization
Sense-making capability of organizations and members (OV5) (Marra 2004)	Project participants feel a high level of trust and credibility when communicating openly and truthful Employees have a good interpersonal relationship
Skill capability of members of the organization (OV6) (Lewin and Regine 2001)	Employees are capable of effectively conveying their opinions and communicate project knowledge and information

(i.e., social relations) are even more important variables that can affect the efficiency of economic activities (Granovetter 1973; Uzzi 1997).

An important affirmation of the chaos and complexity theory is "that the stronger the connections between the diversity of elements comprising a system, the more capable the system will be of sustaining itself when far from equilibrium" (Fitzgerald 1996, p. 29). At the edge of chaos, when a lot of information flows between the diverse agents of a system and when the agents are richly connected, the system employs "a self-correcting mechanism" (Paraskevas 2006) using cognitive processes to perceive weak links among the agents and restructure agent connectivity to trigger the structure change. Rich connections are characterized by repeated interactions, trust, reciprocity, mutually confiding, collaboration, low level of scapegoat, personal informal ties, and so on (Nooteboom et al. 2000; Stacey 1996; Uzzi 1997). Coordination and collaboration can be obtained by a number of means including hierarchy, formalization, and centralization within the organizational context (Burton and Obel 2004).

Links or ties in a social network are considered to be the bridges upon which knowledge sharing occur. The strength or weakness of the tie determines what type of knowledge is shared (Fernie et al. 2003). Strong ties, identified by high level of trust, lengthy timeframes, and close relationships, are ideal for the sharing of tacit, complex knowledge (Uzzi 1997). In contrast, weak ties limit the exchange of knowledge and even information (Fernie et al. 2003). Communication in social networks is based on the sharing of strategic information to help network participants to make better decisions and become more responsive to change. The ability to change and strategically manage an organization will lie in the challenges of relationship management, not in changing the structures or functions of individuals, or of neatly packaged strategic formulations (Fitzgerald 1996). A flexible communication management process and procedures will help to enhance the relationship management and allow for open communication. Therefore, for the communication management, the tighter the relationship among project members along the project's life cycle, the higher the degree of mutual trust among the project members and the richer the connection will be. Strong effective informal ties increase feeling of security and can thus make a group of people more likely to change (Krackhardt 1992).

However, too much interdependence between project participants can result in an increased communication load, more rework, even decreased quality, and missed deadline. Over tightness or over embededness of the relationship may have a negative impact on the performance of a network. If ties between participants within the network are too strong, feelings of obligation and friendship may be obstacles sometimes toward the resolutions of certain problems. Indeed rich connections are extremely vulnerable to over-burden relationship and even perceived defection. Axelrod (1997) has shown that even in a simple game such as tit-for-tat, any misunderstanding or perception that one of the members of a network is cheating or defecting will likely spoil the relationship, sometimes permanently, and it will provoke retaliations from the members who feel that they have been cheated. Hence, transparency of intent and straightforward actions will be helpful to prevent misunderstanding. Good communication skills to discourse and express own opinion is highly important.

A large body of research on social network also explores the relations between interdependent actors (Wasserman and Faust 1994). Much research on social networks has focused on the "specification of the principles of exchange and power that apply to different kinds of network structures. In particular, attention has been focused on the relationship between types of exchange connections and the distribution of power and dependence among actors in various network structures" (Cook and Whitmeyer 1992, p, 113). There are similarities to the edge of chaos idea from complexity theory in some network exchange studies. For example, Markovsky et al. (1993) used both computer simulations and laboratory experiments to look at the capacity of actors in different positions in a network to accumulate resources. They found that the most interesting dynamics occur in networks having moderate density of connections among actors. While simple patterns were observed in both relatively dense networks and relatively sparse networks, moderately dense networks yield complex phenomena such as counter-intuitive exchange patterns and decompositions of the network (Markovsky 1998, p. 3).

The process and the building of relationships between human objects are vital, and development and maintenance of these relationships are of more importance than the objects themselves. Because of the interdependency of systems with the environment, relationships actually give meaning to the entities and processes and meaning is not situated within the entities or processes themselves (McDaniel 1997, p. 24). However, understanding the complex network of connections among individuals and the pattern of communication and influence is difficult and can never guarantee a 100 % completion rate. Wheatley (1994) also contended that there is interdependence among different subsystems in an organization, and this interdependency suggests that all the subsystems should take part in the process of the system. Gulch (1993) argued for respectful interaction under crisis situation, based on the principles of trust, honesty, and respect in communication interaction. Gulch (1993) also discussed that this triangle of trust, honesty, and respect is conspicuously missing in several well-documented disasters in which faulty interaction processes lead to increased fear, and diminished communication.

Granovetter (1973) demonstrated that knowledge and information, which circulates among members of a group via strong ties, will be locked within the group, thus impeding their dissemination across groups. The ties are strong when a group of people spend much time together, are emotionally involved, and mutually confiding, and provide reciprocal services (Stacey 1996). Particularly during a crisis, groups perform important functions such as enabling people to solve tasks of much greater magnitude and complexity than they would be able to handle alone (Loosemore 2000). Hence an awareness of group formation is very important for crisis management. It can be used by the crisis manager to prevent communication breakdowns.

However, the effect of too strong ties is to bind people together, making it more likely that behavior will be repetitive and uniform. Hornstein (1986) found that groups have a capacity for producing a special, perniciously subtle tyranny that can damage communications, slow down decision making, produce compromise decisions, and suppress creativity and innovation. This also refers to the effect of "groupthink". Loosemore (2000) illustrated the potential danger of groupthink in the Challenger Space Shuttle disaster when, on the evening before the launch, engineers, with full knowledge that the expected temperature at the time of launch would be below safety levels, were pressured by peers to sanction the launch. Hence it would seem that during a crisis, when effective communication, open-mindedness, creativity, and flexibility are most valuable, groups could be at their most vulnerable. The potential damage and danger of groupthink should also be considered by the crisis managers during a crisis to prevent communication breakdowns.

The above discussion is summarized in Table 4.2.

4.3.4 CP3: Levels of Diversity Among Agents of the System

Holland (1995) refers to diversity as one of the properties universal to Complex Adaptive Systems. Elements within a system can exhibit a variety of attributes, and these attributes can change and expand over time. According to evolutionary biologist Simon Levin (1998), the maintenance of diversity and individuality of components implies the generation of perpetual novelty, and far-from-equilibrium dynamics. A low level of diversity among agents' schemas is usually characterized by conforming members who operate in the legitimate system and do not challenge the organization's established rules. Here schemas of agents are dictated by agents' mental model (the conscious and unconscious rules and patterns of behavior that individuals typically follow) and the organizational institutions that constitute the framework within which the agents may behave and which create the culture of an organization (Stacey 1996). An organization with a low level of diversity among agents' schemas will remain in the equilibrium state. A high level of diversity among agents' schemas is characterized by members who are able to challenge the pre-established institutional framework imposed by an organization.

Table 4.2 Measurable OV affecting CP2

Organization variables	Operationalization
Organizational structure (OV1) (Burton and Obel 2004)	Hierarchy structure (Burton and Obel 2004)
	Bureaucracy structure (Burton and Obel 2004)
	Decentralized structure (Burton and Obel 2004)
	Formalization administrative structure (Burton and Obel 2004)
Organizational culture (OV2) (Nooteboom et al. 2000; Uzzi 1997)	Project participants are willing to knowledge exchange and strategic information sharing (Fernie et al. 2003; Fitzgerald 1996;)
	Collaboration and cooperation between project participants (Nooteboom et al. 2000; Uzzi 1997)
Information technologies capability and information management system (OV3) (Elliott et al. 2000; Fernie et al. 2003; Segel 2000)	Web-based technology is used to support project communication (Hapgood 2004; Ravichandran 2005)
	Project members meet face-to-face or via teleconference on a regular basis (Burton and Obel 2004)
	Feedback and response to the issues are prompt in a timely fashion (Burton and Obel 2004)
	Innovative integration information system is adopted (Hapgood 2004; Ravichandran 2005)
	The same information is available to the appropriate parties (Burton and Obel 2004)
Style of leadership and management (OV4) (Fitzgerald 1996)	A flexible communication management process and procedures to allow for open communication
	A rigid and centralized communication management process and procedures
Sense-making capability of organizations and members (OV5) (Stacey 1996; Uzzi 1997)	Project participants feel a high level of trust when communicating openly and truthful (Gulch 1993; Uzzi 1997)
	Project participants consider providing reciprocal service (Nooteboom et al. 2000; Stacey 1996; Uzzi 1997)
	Project participants show respect to one another, and hold accountable to each other (Gulch 1993)
	Project participants demonstrate transparency of intent and straightforward actions (Axelrod 1997)
	Distortion and misunderstanding of information is common during work (Axelrod 1997)
	Group thinking is a common phenomenon during work (Hornstein 1986; Loosemore 2000)
	Project participants have a very strong feeling of obligation and friendship (strong personal ties) (Nooteboom et al. 2000; Uzzi 1997)
	Problems can be recognized in the very first sign and dealt with quickly, without transfer of blame to those who are not responsible (low level of scapegoat) (Nooteboom et al. 2000; Uzzi 1997)
Skill capability of members of the organization (OV6) (Axelrod 1997)	Employees possess good communication skills to discourse and convey their opinions

In such a way an organization that presents too much diversity among its agents' schemas will tend to move toward chaos (Stacey 1996). Ideally, an organization should have a certain level of diversity which is enough to allow the members of the organization to question the existing state, but not so high as to inhibit the maintenance of some level of cohesion.

As applied to the communication practices, when two groups of people with different discipline backgrounds use different vocabularies and concepts but in talking to each other, trying to understand each other's ways of talking, new meaning may be generated (Fonseca 1998). Diversity arises in misunderstanding and in the cross-fertilization of concepts through interaction between different patterns of conversation (Stacey 2000). If there is little misunderstanding between people forming a group with well-established concepts and ways of talking to each other, their conversations are likely to be repetitive. If there is too much misunderstanding between people drawn from very many disparate groups then there is the disintegration of communication. Therefore, the conditions for creative, free-flowing conversations lie in some critical range between these extremes. Project managers should not seek unanimous control and command but should work toward diversity and different voices in order to keep the system creative and on the edge of chaos (Stroh 1998). They should facilitate a climate of constant change, conflict, and diversity where the communication management function becomes the strange attractor keeping the organization from entropy and instilling a culture of positive chaos (Stroh 1998).

Human communication generates both meaning and social structures. Social structures consist of beliefs, values, norms, or practices, which are associated with a given culture. Capra (2002, p. 87) explains: "Culture arises from a complex, highly non-linear dynamic. It is created by a social network involving multiple feedback loops through which values, beliefs and rules of conduct are continually communicated, modified and sustained. It emerges from a network of communications among individuals and, as it emerges, it produces the constraints on their actions. In other words, the social structures, or rules of behavior, that constrain the actions of individuals are produced and continually reinforced by their own network of communications". Within the context of communication system of a construction project organization, it may seem that a very high level of differentiation and inherent in construction professional specialization and division of labor (Baccarini 1996) are the possible factors affecting communication effectiveness.

In system terminology, such specialization is called differentiation (Lawrence and Lorsch 1967). There is a built-in conflict between different professional and occupational cultures and that they have learned their assumptions from the general occupational environment within which they exist (Schein 1996). These differences imply a likelihood of conflict at the interface level where the human interaction elements come into play. An integration knowledge information system can facilitate effective information sharing and communication between individuals with different professional background.

Therefore it is a function of project management to align various professional cultures to the project's core values. As reported by Fellows and Seymour (2002), occupational and organizational differences, how they affect receptivity to new practices and technologies and interfirm collaboration is one of the two main focuses of culture research in the construction industry. Low and Shi (2001) presented a study on cross-cultural working experience which showed *inter alia* that mismanaging cultural differences rendered otherwise successful managers and organizations ineffective and frustrated when working across cultures. Scott (1998) argued that the idea in the popular organization literature of benefits of creating a "strong" culture, which aims to sustain a "commitment to something larger than self", can also develop into an authoritarian system that is subject to abuse.

Managing construction projects is about harnessing the various cultural forces at play in construction projects to the benefit of projects and therefore an adequate understanding of the organization's culture, beliefs, and value system is important for project manager to harness them, other than fight them (Walker 2006). In terms of crisis response communication management, a proper crisis management plan that consists of a formalized set of procedures can improve response times by collecting information in advance (Coombs 2007) to direct and guide the communication actions and behavior in a structured way, and allow the organization to take proactive control of the crisis situation (Kash and Darling 1998).

As learnt from complexity theory, excessive diversity leads to instability and results in conflicts and misunderstanding among project members. The development of a shared vision is important for smooth network operations (Javenpaa and Ives 1994). In terms of crisis response management, Paraskevas (2006) suggested that a common goal should serve as an internal model for the organization's self-organizing processes and should be co-created with all the actors involved. This co-creation of a crisis culture can be propagated to permeate the entire organization. In order to develop such a vision and goal alignment, individuals in the communication network will need to learn from each other's behavior and adapt to the behavior of others. Discourse is an important concept that refers to the use of language in communication by forming structures and conveying meanings (Holtzhausen 2000). Hence possession of good communication skills is of importance for the individuals to be capable of coping and resolving differences in opinions. Holtzhausen (2000) explained that meaning is not formed through language itself but by the debate or discourse of different points of view as well as in the ways knowledge is structured. Discourse thus creates and structures ideas, beliefs, and ideology.

During the crisis response stage, diverse agents from several different parts of the organization communicate and conduct one or more crisis response tasks. All parties contribute in the crisis situation through their interaction at a local level. In practice, different alignments will appear and as both the risk and responsibilities are shared (Comfort 1999), they will start competing for more resources. Hence a shared understanding which means that the different actors are fully aware of the risks and benefits entailed in the development of the project, and that they are fully aware of their rights and obligations as well as those of the other parties, is essential for the development and maintenance of a cooperative relationship (Ring and Van de Ven 1994).

Crisis often creates or exacerbates tensions of conflict and division as crisis stakeholders offer competing interpretations of the blame and responsibility over compensation. People are likely to express contradictory views about the criteria of project performance and the value of the outcomes. This can be illustrated by addressing the concept of differentiation and diversity in the conventional organizational studies and social network (Burton and Obel 2004; Galbraith 1973; Hall 1991). Horizontal differentiation refers to specialization within an organization, and vertical differentiation relates to the depth of the hierarchy of the organizational structure (Burton and Obel 2004).

The traditional view of conflict in the construction industry is that of a disruptive force that should be avoided and eliminated at all cost (Loosemore 2000). The conflicts of interests of different parties that inevitably exist in construction organizations complicate the people's behavior especially within the changing environment. However, the potential for conflict during a crisis should not be seen as entirely destructive. Comfort (1996) suggested that there is a need for sufficient structure to hold and exchange valid information that will support different agents' actions toward the common goal and processes of self-organization where informed agents initiate action, but adjust their action to that of others operating toward the same goal in accordance with changing needs. A well-managed conflict can force a more thorough investigation of a wider range of crisis solutions and can change people's attitude to provide the foundations for constructive conflict management.

The above discussion is summarized in Table 4.3.

4.3.5 CP4: Contained Anxiety and Tension

Stacey (2001, p. 391) contended that the "good enough holding" of anxiety is an essential condition for the free-flowing conversational dynamics that is the analog of the edge of chaos. "Anxiety is thus an inevitable feature of mental life at the edge of chaos; the ability to bear that anxiety is a prerequisite for dwelling there. To obtain such anxiety, an individual requires a strong ego structure and a good enough 'holding' environment. That is, an individual needs to be able to hold ambiguities and paradoxes without being overwhelmed by anxiety" (Stacey 1996, p. 188).

Stacey (1996) also suggested that an individual mind is a nonlinear feedback network that is part of a larger nonlinear feedback network of interacting minds. To demonstrate this point, Stacey (1996) borrowed from psychoanalytic studies to understand the organizational effects of anxiety. First, there are the defenses people who use to avoid feeling anxious. These may take the form of structures and procedures having the ostensible purpose of enabling some rational task, but actually operating as defenses. For example, people may prepare forecasts of

Table 4.3 Measurable OV affecting CP3

Organization variables	Operationalization
Organizational structure (OV1) (Burton and Obel 2004; Galbraith 1973; Hall 1991)	Simple hierarchy structure (Burton and Obel 2004)
	Bureaucracy structure (Burton and Obel 2004)
	Ad hoc structure (Burton and Obel 2004)
	Routine administrative structure (Burton and Obel 2004)
Organizational culture (OV2) (Capra 2002; Stroh 1998)	Understand and commit to the organization's vision (creating strong culture) (Javenpaa and Ives 1994; Scott 1998)
	Employees share common crisis goal and crisis prepared mindset (Paraskervas 2006)
	Employees have an adequate understanding of organization's belief and value system (Capra 2002; Walker 2006)
	Facilitate a climate of constant change, conflict and diversity (Stroh 1998)
	Encourage discourse of different points of view (Holtzhausen 2000)
Information technologies capability and information management system (OV3) (Baccarini 1996)	An innovative integrated information system is adopted
Style of leadership and management (OV4) (Comfort 1996; Loosemore 2000)	A constructive conflict management for a well-managed conflict (Comfort 1996; Loosemore 2000)
	A conflict avoidance and elimination preference (Comfort 1996)
	A proper crisis management mechanism and system (Coombs 2007; Kash and Darling 1998)
Sense-making capability of organizations and members (OV5) (Ring and Van de Ven 1994)	Employees are fully aware of their risks and obligations as well as others (Ring and Van de Ven 1994)
Skill capability of members of the organization (OV6) (Holtzhausen 2000)	Employees possess good communication skills to discourse and convey their opinions (Holtzhausen 2000)
	Employees is capability of coping and resolving differences in opinions (Holtzhausen 2000)

future states that are impossible to predict and develop strategic plans on the basis of these forecasts. Such plans may then just create a sense of certainty to defend people against the anxiety of feeling uncertain. The result is stable, repetitive conversational dynamics around strategies that are simply a continuation of what is already being done. Unfortunately using this defensive mechanism strategies increase the resistance to change.

The psychoanalytic concept of "good enough holding" presents the conditions that people are able to hold the simultaneous excitement and anxiety of conversations that test the boundary of what they know. It is a quality of the themes organizing the experience of relating. When these take the form of trusting interaction, they are themselves then forms of good enough holding. In another words, when the quality of relating is characterized by mutual trust, conversation can take free-flowing forms. Also the quality of power relations is closely related to the "good enough holding" of anxiety. Themes organizing that relate between people may be highly constraining so that power relations have the qualities of force, authoritarianism, and so forth. The responses that these qualities evoke are either submission or rebellion. The former produces highly repetitive, stable conversational pattern, while the latter produces disintegration in communication.

Stacey (1996) suggested that holding organizational anxiety requires not only internal container, which is provided by a culture of trust and particular patterns of power use, but also external containers, which must be provided by the industry and society to which the organization belongs. Organizations turn out to be creative and innovative only when they engage the creative processes of other organizations and so amplify the schema changes across industries, economies, and societies. Thus, the manner in which others in a society treat an organization affects the level of anxiety the organization experiences. When an industry and society provides a supportive emotional environment for an organization, its members are able to hold higher level of anxiety and therefore may be more creative.

It has been demonstrated that very high levels of integration between diverse professionals and experts working with one another can generate anxiety (Allen 1994). Allen (1994) showed that the implementation of information systems designed to increase activity integration among different experts or professionals by providing real-time information increases the level of mutual control. Such systems force professionals to look for problems outside their own domain and to coordinate efforts in response to others' unexpected change.

Crises are potentially serious events that require inventive solutions under extreme pressures in a timely fashion (Loosemore 1998a). This ensures that those affected feel a certain degree of tension and anxiety. Some argue that such feelings will produce a determination that is important to the efficient resolution of problems, while others point to them increasing suspicion and reducing communication. George (1991) pointed out that stress is a generic problem that poses severe threats to crisis management. The stress associated with a crisis arises from the dramatic challenge to previously held views, from the dislocation to social relations and from the physical challenges posed.

In the face of a crisis, victims experience a heightened sense of vulnerability, and their sense-making and rationality are impaired (Weick 1993). The erosion of the individual and collective assumptions and their sense of meaning can probably lead to the ultimate failure outcomes. The sense-making capacity of individuals seeking to understand and contain a crisis influences the effectiveness of the individual's response to a crisis. The implementation of a well-defined crisis management plan is a means to contain individual's anxiety since it aims to push an organization's shadow system toward the stable zone, however too heavily relying on and following mechanically these detailed sequence of tasks and procedures may also contain individual's creativity in response to a crisis. Three means are proposed by which anxiety levels can be controlled, but not avoided, within an organization at the edge of chaos (Stacey 1996). These means are: (1) improving the quality of the relationships between the individuals in an organization or group; (2) using self-reflection; and (3) improving the quality of leadership and the way power is exercised.

Good quality of relationships increase the likelihood that the individuals trust and like each other to a reasonable extent, and feel supported by their peers and show positive attitude to the organization that they are willing to raise issues and take risks within the organization. Stacey (1996) considered successful self-reflection renders an individual more capable of accepting facts and feedback and holding the paradoxes of the depressive position. Leadership is a very important means to hold anxiety within an organizational context. Anxiety-containing capacity is a function of the behavior of the leader that has to do with the manner in which power is used and with compassion for the feelings and fears of others in the organization (Stacey 1996). Leaders contain anxiety when they are able to empathize with others and articulate or interpret what they are experiencing (Carr and Shapiro 1995). Leadership includes not only the capacity to contain anxiety for others, but also the ability to provoke and contribute to the double-loop learning process on the others (Stacey 1996). A good leader may also contribute to encourage a sense of purpose especially during a crisis or in a chaotic environment, thus lowering the level of anxiety of the project members. Meaning and sense of purpose help an organization transcend the messiness of its organizational context and provide a point of reference on which individuals can base their actions (Wheatley 2006).

In addition, project managers or leaders should not be "invested in establishing themselves as the ultimate authority"; instead of directing people, they "cultivate conditions where people could self-organize and restructure around the existing issues" (Lewin and Regine 2003, p. 173). Thus, the characteristics of influence rather than control are the key from a complexity perspective, especially necessary to coordinate crisis response. Developing the capability of learning and adaptation will facilitate to create an effective communication environment (Zhong and Low 2008) and help to develop an agile response to a crisis situation.

The above discussion is summarized in Table 4.4.

4.3.6 CP5: Power Differentials

Stacey (1996) stated that when power differentials are extreme with a fixed elite wielding most of the power, and when that power is exercised as force or oppressive authority, an organization's shadow system is driven deep underground and its members operate in fear and will not challenge the authority imposed on them, then the organization will operate in the stable zone. Conversely if power is equally distributed among all the members, the organization may enter into a disintegrated state as no one really exercises that power. Between these two

Table 4.4 Measurable OV affecting CP4

Organization variables	Operationalization
Organizational culture (OV2) (Stacey 1996)	A climate that encourages mutually trusting interaction (Stacy 1996)
Information technologies capability and information management system (OV3) (Allen 1994)	An innovative integration information system is adopted (Allen 1994)
Style of leadership and management (OV4) (Stacey 1996)	Authority control preference leadership style (Lewin and Regine 2003)
	Influencing, motivating and inspiring leadership style (Lewin and Regine 2003)
	Encourage a defined sense of purpose (Wheatley 2006)
	Strictly follow the procedures of the crisis management/communication plan (Stacey 1996)
Sense-making capability of organizations and members (OV5) (Stacey 1996; Weick 1993)	Development of defensive mechanism strategies to avoid anxiety (Stacey 1996)
	Capable of handling the external influence under dynamic environment, e.g., industry and society (Stacey 1996)
	Employees hold self-reflection attitude to be willing to engage in creative double-loop learning process (Stacey 1996)
	Employees tend to withhold negative information to protect themselves and demonstrate a high level of resistance to change (Stacey 1996)
	Project participants are confident and show a positive attitude to the project organization (Stacey 1996)
	Employees are willing to raise issues and take risks (Stacey 1996)
	Employees are willing to accept the facts and constructive feedback (Stacey 1996)
	Employees have a good interpersonal relationship (Stacey 1996)
Skill capability of members of the organization (OV6) (Stacey 1996)	Employees possess good problem-solving and analytical skills to handle unforeseen circumstances (Loosemore 1998a)
	Employees possess capability to cope with pressure inherent in crisis situations (Loosemore 1998a)
	Employees possess capability to hold paradox and ambiguities (Stacey 1996)

extremes, a critical point is reached where anxiety is contained through a clear hierarchical structure and freedom to express opinions is achieved (Stacey 1996). The manner and extent to which roles, power, and responsibilities are delegated, controlled, and coordinated are determined by organizational structure (Business Dictionary 2009). In a centralized structure, the decision-making power is concentrated in the top layer of the management and tight control is exercised. In a decentralized structure, the decision-making power is distributed and the autonomy is exercised in varying degrees.

As applied to the communication process, communication is a complex process of relating (Stacey 2001), which involves both words that are spoken and the response that they elicit (Kauffman 1993), and centered around the role of language that is simultaneously used for conversation and to negotiate social status and power relationships (Stacey 2003). These complex responsive processes are fundamentally conversational in nature, forming and being formed by power relations (Stacey 2000). The maintenance of human relationships imposes constraints. Power is a constraint that excludes some communicative actions and includes others. The process of communicative interaction reproduces and transforms themes of emergent patterns of collaboration, and at the same time reproduces and transforms themes to do with inclusion and exclusion, or power (Streatfield 2001). Suchman (2000) argued for a view of construction projects as "persuasive performances", illustrating how power operates through conversations and artifacts through which project-based work is being organized and performed in real-time in a specific context.

Ragins (1995) also defined power as the influence of one person over others, stemming from an individual characteristic, an interpersonal relationship, a position in an organization, or from membership in a societal group. These perspectives on power reflect individual, interpersonal, organizational, and societal levels of analysis. From individual power perspective, there can be two further kinds of power: expert power and referent power. Expert power arises from skill, professional knowledge, and also increasingly information, and depends largely on the personal attributes of the individual, such as expertise and charisma (Greiner and Schein 1988). Finkelstein (1992) believed that the ability of top managers to deal with environmental contingencies and contribute to organizational success is an important source of power. Newcombe (1996) argued that building expert information power is an important aspect of relationship interactions in construction projects. Referent power is the influence that the project managers or leaders exercise because people believe in them and their ideas, and also the project mangers or leaders are confident in the organization and find delegation efficient in their own use of time (Burton and Obel 2004). Finkelstein (1992) referred to this as prestige power. The strength gained through expert or referent power will be increased by an appropriate personality (Robbins 2005).

From organizational perspective views, power resources are defined as control over persons, information, and organizational resources which also involve the development of authority, credibility, and perceived expertise (Hinnings et al. 1974; Pfeefer 1981). In the communication network, each participant holds a position of relative power and influence over the others within the network. Individuals who control essential resources or can form coalitions have strong power (Coombs 2007).

Under a sociological perspective, power is viewed as fluid relationship between groups (Blalock 1989) in the larger ecosystem such as the construction industry or society. Power relationships among groups may be symmetrical or asymmetrical, whereby one group dominates another group and has more power resources (i.e., economic resources, prestige, natural resources, authority) (Blalock 1989). The fluid and dynamic properties of intergroup power relations may result in subtle or dramatic shifts in power relations among groups in organizations. Ragins (1995) pointed out that groups with power are interested in maintaining their influence and resources, and may do so by supporting policies, practices, and prescription that exclude other groups from power. Ragins (1995) argued that society and those in the organization with power substantially shape power relationships among groups in organizations (Ragins 1995, p. 96). The nature of power and influence, the resources of this power and the way in which it is used to contribute to, or manipulate cooperative relationships underpin all procurement strategies and the relationships that develop from these.

Pfeffer (1992) recognized the link between power and interdependency in organizations. With little or no interdependency there is no need for power as there is no situation in which dependency in the basis of power occurs. Similarly, Pfeffer (1992) believed that when Interdependency is high the motivation to work together is also high and that if this incentive is ignored, the organization is likely to fail. Hence, it is in condition of moderate interdependence that power is more frequently used. In terms of the communication network, the interdependency influences the distribution of power between members of the network. Thus, understanding the patterns of interdependency and power structure is critical.

Thorelli (1986) specified five interdependent sources of power in a network that influence power differentials. These sources of power are: economic capital, technology, expertise, trust,and legitimacy. Within the construction project, economic power may be manifested through the client's decision over budget allocation and the contractor's ability to remain within budget. Technological superiority may be expressed through a discipline's capability to innovate, or the amount and type of knowledge owned. Expertise is reflected in professionals' capability to apply their knowledge and expertise to resolve project issues. Trust can also be a source of power when it is reciprocal and it forces participants to cooperate. Building a collaborative relationship between participants through developing mutually agreed goals and objectives is a task easier preached than practice. Legitimacy as a source of power can originate from the type of contractual mechanism used between project participants.

Construction management is based on the modern management principle of empowerment or power equalization and reflects the trend toward a more pluralistic society (Newcombe 1996). The empowerment was defined as the delegation of authority and the increased involvement of lower level employees in the control and distribution of resources (Burke 1986). Newcombe (1996) argued that when ultimate authority rests with the client, the encouragement of participation by all the parties in decision making, coupled with the democratic distribution of power provides a radically new basis for conducting construction process. The

Table 4.5 Measurable OV affecting CP5

Organization variables	Operationalization
Organizational structure (OV1) (Stacey 1996)	Clear hierarchy structure (Stacey 1996) Centralized bureaucracy structure (BD 2009) Decentralized structure (BD 2009)
Organizational culture (OV2) (Thorelli 1986)	Open-mindedness environment for building mutually trust relationship (Thorelli 1986) Learning environment to constantly acquire and communicate knowledge within the organization (Coombs 2007; Zhong and Low 2009)
Information technologies capability and information management system (OV3) (Thorelli 1986)	Technology capability to innovate (Thorelli 1986)
Style of leadership and management (OV4) (Ragins 1995)	Manager's expertise to deal with crisis contingency (Finkelstein 1992; Robbins 2005) Manager's ability to influence over others (Finkelstein 1992; Robbins 2005) Control of authority leadership style (centralized) (Hinnings et al. 1974; Pfeefer 1981) When a critical decision must be made, there is always someone in the project who has the power to delegate decision (Burton and Obel 2004) Delegation of authority leadership style (empowerment) (Burke 1986; Hinnings et al. 1974; Pfeefer 1981) Participative decision-making process (democratic) (Newcombe 1996) Most decision are made through consensus (democratic) (Newcombe 1996)
Sense-making capability of organizations and individuals (OV5) (Thorelli 1986)	Employees hold trust and credibility towards each other (Thorelli 1986)
Skill capability of members of the organization (OV6) (Ragins 1995)	Some of the employees possess competent technical skills and expertise that would be difficult to replace (Greiner and Schein 1988; Ragins 1995)

clients should realize that the criteria used to select consultants, contractors, and the form of contract may be less important than the approaches to power structures.

The involvement of a party or a person in the management of crisis is determined by the extent to which their vested interests are threatened. It is usually the case that the weak tended to be excluded from the process and the powerful dominated it (Loosemore 2000). Traditionally, the emphasis has been on the power elite within an organization regaining control during or after a crisis, and the success of crisis communication efforts has traditionally been evaluated on how quickly that control has been regained (Coombs 2007). However, from a complexity point of view, even the most experienced professionals cannot fully control

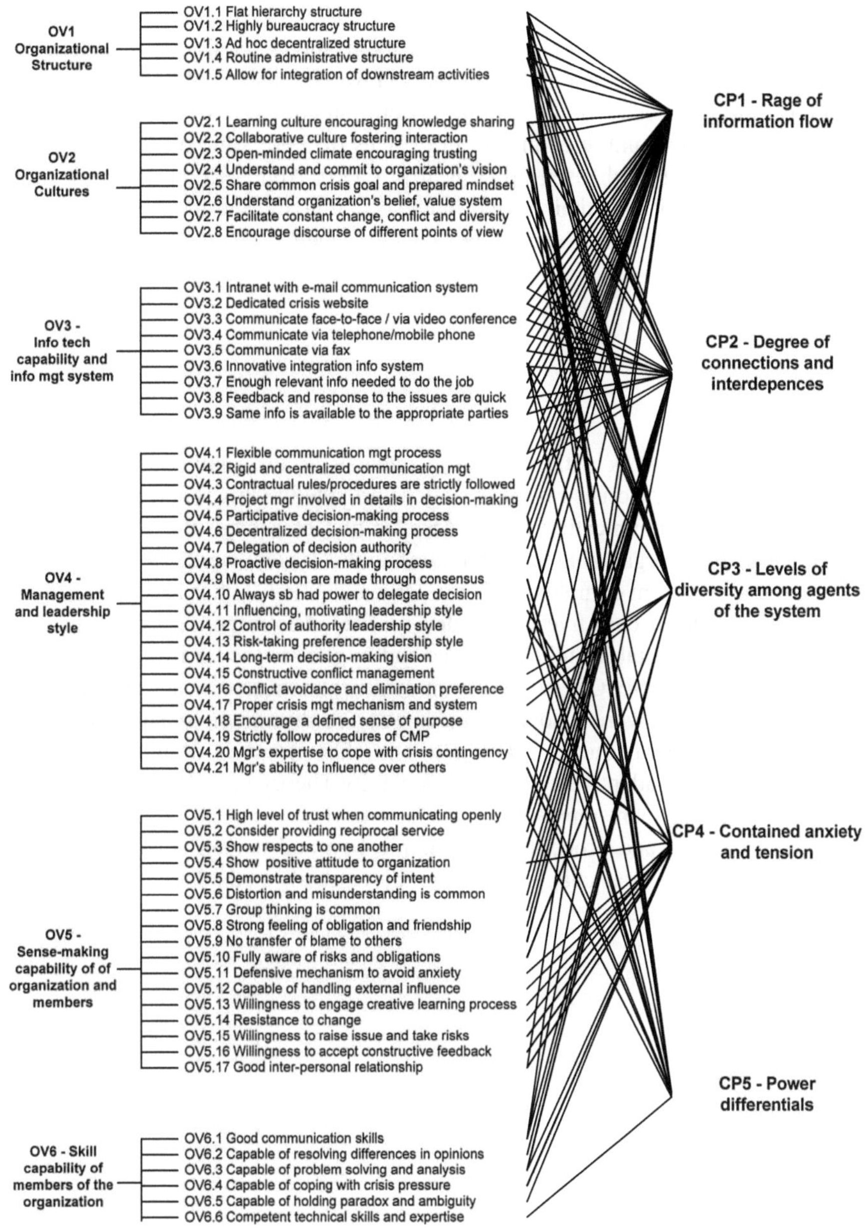

Fig. 4.4 Summary of OV and CP

the crisis situation as the progression of multiple minor problems can aggravate the organizational crisis, but they can learn to work with adeptly (Zhong and Low 2009). The consequent need is a learning process within the organizations by which knowledge is constantly communicated and acquired.

Health and Millar (2004) suggested that a crisis can be viewed as "a struggle for control" and Coombs (1999) advocated establishing a "crisis control center" to facilitate "crisis containment and recovery". However, instead of focusing on how to protect those already in power, the efforts should also be directed toward resisting domination and attending to the voices of those who are disemboweled and marginalized (Holtzhausen 2002; Martin 1990). Appreciation of people with different power positions, their varied interests and goals, and the communication patterns which serve them should be acknowledged. The above discussion is summarized in Table 4.5.

4.4 Summary

A theoretical framework within which to understand and analyze the communication system during a crisis response stage is provided in this chapter. A model of crisis response communication as a CAS was first presented; the characteristics and process of self-organizing and adaptation in this system were briefly reviewed. Later the findings from complexity theory, organizational studies and management were combined to discuss five CP needed for the crisis response communication system to exit the unstable and chaos zone and migrate to the far from equivalence or the edge of chaos. Six OV derived from organizational studies and management to influence the CP were also explained, and the discussion is summarized in Fig. 4.4. As a general theory relating to the characteristics of nonlinear, self-organizing complex adaptive systems, Complexity Theory linked to organizational studies and management represents a useful analytical framework to explain the underlying patterns and process of communication and human behavior systems during a crisis. The next chapter presents the research method for the empirical investigation.

Chapter 5
Research Methodolgy

5.1 Introduction

This chapter discusses the research methodology of this study. An appropriate research design is essential for achieving the research objectives. The research is an empirical study with a combination of the quantitative and qualitative approaches through questionnaire survey, interviews, and case-study analysis. Quantitative studies serve as a means to identify and evaluate the significant factors/variables which influence the crisis response communication system in order to creatively respond to unforeseen events and overcome adaptive challenges during a crisis. The qualitative studies, on the other hand, serve as a means of illustrating and further analyzing the findings from the quantitative studies, in order to gain a deeper understanding of the various aspects of the organizational communication system in response to a crisis. There are advantages to adopting multiple research methods to examine the same aspects of a problem. The weakness of one method will compensate the counter-balancing strengths of another, and thus the validity and reliability of the research is strengthened (Jick 1979; Lincoln and Cuba 1985).

5.2 Research Methodology

The flow of methodology is shown in Fig. 5.1. This research was conducted in four major phases: (1) research model development; (2) survey instrument development; (3) case-study development; and (4) research validation phases.

Z. Ying and L. Sui Pheng, *Project Communication Management in Complex Environments*,
DOI: 10.1007/978-981-4560-64-1_5, © Springer Science+Business Media Singapore 2014

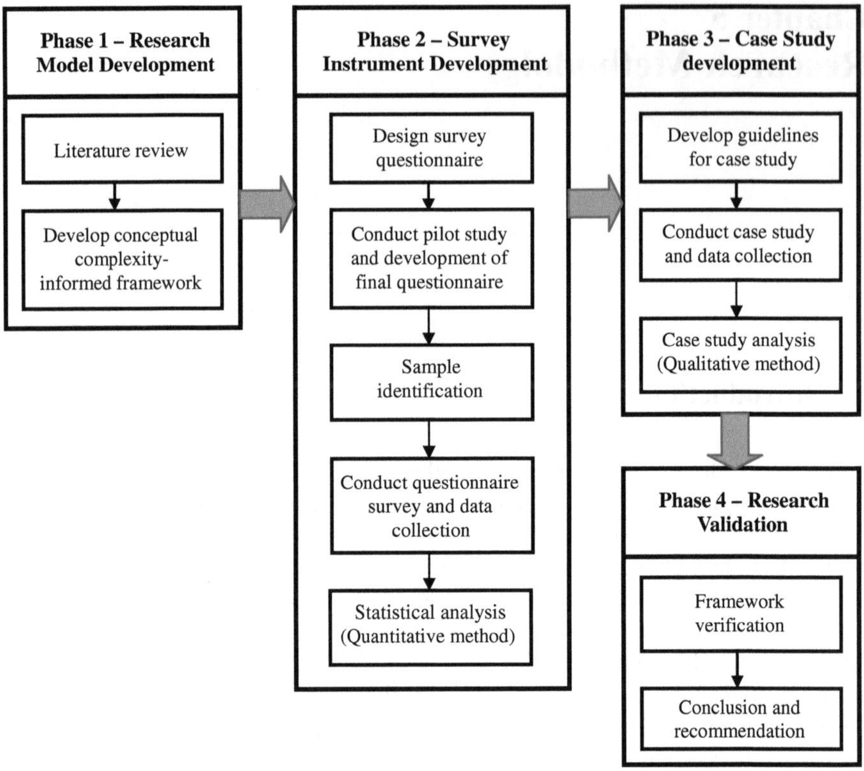

Fig. 5.1 Research methodology

5.3 Research Model Development

An essential objective of the comprehensive literature review is to acquire the comprehensive knowledge about the crisis response communication management and complexity theory through various research works completed in the past. For an understanding of the influence of organizational variables on the control parameters derived from complexity theory, analytical review of organizational studies and management literature from strategic organizational design, social network, learning organization theory, and knowledge management was carried out. This literature is available in books, journals, and professional magazines. A comprehensive catalog was developed from the literature which would serve as the guideline for further research in this study (see Fig. 5.2).

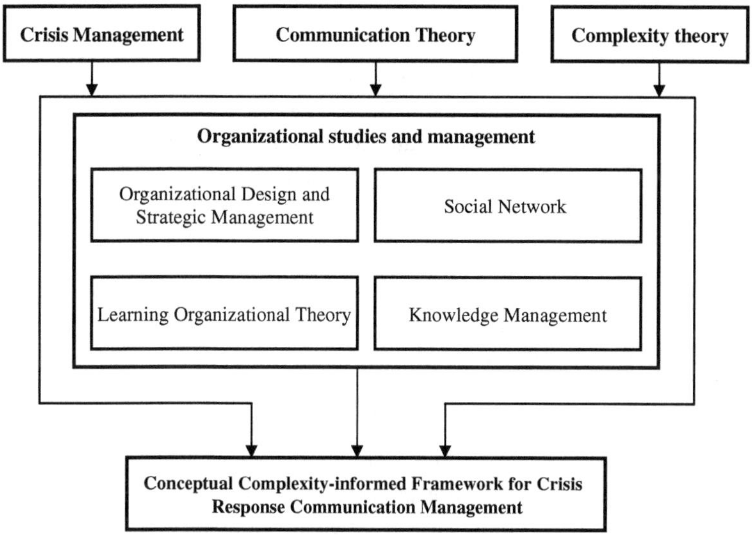

Fig. 5.2 Literature review catalog

5.4 Survey Instrument Development

In this research, a survey design was selected in the survey instrument development phase for its abilities to provide a relatively efficient approach to (i) obtain information from the targeted sample (Tan 2002), and (ii) generalize the research findings based on the sample involved (Gill and Johnson 1997).

For the survey development phase, a self-administered postal questionnaire was used as the main data collection tool. Generally, it requires respondents to complete the posted questionnaire and return their replies through post using the enclosed self-addressed and prepaid postage envelop. The advantages of selecting this technique are as follows:

(i) It provides greater geographical flexibility as compared to face-to-face interviews (Tan 2002);
(ii) It allows data to be collected from a diverse sample at a relatively low cost as compared to other methods (Wang and Li 2007); and
(iii) It allows respondents to have sufficient time to think about the questions and respond at their own convenience (Tan 2002).

However, a self-administered postal questionnaire does have the disadvantages of: (i) incompleteness of returned questionnaire; (ii) results may be biased if questions are misunderstood; and (iii) poor response rate (Tan 2002).

In an effort to overcome shortcomings of the self-administered postal questionnaire, face-to-face interviews were also conducted to ensure that all questions were answered. The use of face-to-face interviews could also be advantageous in that the respondents could, if necessary, fully probed the meaning of the questions and have a chance to clarify any doubts with the researcher immediately. This approach also allowed general discussion and peripheral comments to be noted to add supporting contextual evidence (Walker 1997).

To further improve the response rate, follow-up telephone calls were made to the remaining firms. The purpose was to explain the aims of the survey and to assure them that all information provided would be treated in strictest confidentiality and that their names and organization's details would be kept anonymous.

5.4.1 Survey Questionnaire Design

Through an intensive literature review, the organizational variables which affect the control level of crisis response communication system of a construction organization were identified. These provided the basis for the formulation of a questionnaire. A draft questionnaire is given in Appendix B Tentative Survey Questionnaire. The questionnaire design was divided into three sections.

Section A was included to collect general information about the respondents on their firm's business nature as well as the general particulars pertinent to the project in which they have been involved, which was affected by the Sichuan earthquake crisis.

Section B was focused on the statements relating to the organizational variables that influence the level of control parameters and thus indicate whether the crisis response communication system is able to overcome the adaptive challenge in the face of a crisis. These statements are the hypothesized variables which were identified in Chap. 4. Respondents were asked to rate each of the variables based on the level of importance according to their professional judgment on a given five-point Likert scale, where 1 represented "*least important*", 2 for "*less important*", 3 for "*important, good to have*", 4 for "*more important*", and 5 represented "*most important*".

Section C requested the respondents to provide their personal particulars including their position and working experience in the construction industry.

A structured questionnaire was adopted in that a "closed" format of questions were asked in the same order and identical questions were posed to all respondents. The selection of structured survey questionnaires over other types was to ensure that the survey questions were standardized in a way that they were presented exactly with the same wordings and order to all respondents. As a result, the respondents were replying to the same questions and thereby the results were comparable and facilitated subsequent analysis.

5.4.2 Pilot Study

Walker (1997) suggested that pilot studies are considered as a useful tool in providing a focus mechanism to establish the research direction more clearly. Kometa (1995) indicated that the purpose of a pilot survey is to test the questionnaire to ensure that it is coherent and comprehensive, and to ensure that meaningful data analysis could be carried out subsequently. After the pilot survey, necessary amendments such as omissions of repetitive questions, and reconstruction of the sentences were made to improve the questionnaire design.

In the pilot study, a face-to-face semi-structured interview technique was adopted because it enabled the respondents to ask for clarification of the questions and allowed instant feedback. The semi-structured approach facilitated more flexibility for the interviewer to modify or provide additional comments on the questions. A small number of industry practitioners were interviewed for the pilot survey purpose. After that, the questionnaire was adjusted and modified accordingly.

5.4.3 Sampling Strategy

A sufficiently large sample of construction firms needed to be identified to enable meaningful statistical analysis of data groups to be undertaken. The research study was restricted to the construction firms who operated building projects in the metropolitan area of Chengdu city, the capital of Sichuan province, which is 80 km away from the earthquake epicenter Wenchuan County. These projects represented the typical projects under construction which were affected by the earthquake crisis to the extent that they have not been completely destroyed or collapsed.

The targeted groups of construction firms in Chengdu were selected under the general building contractor's category from the Sichuan Construction Bureau's General Contractor Registry. It provided a complete list of registered general contractors in Sichuan. Details of their building projects were retrieved from the construction permit/license registry system which was a publicly-available database officially maintained by the Chengdu local government agent, the Chengdu Construction Committee (CDCC). In this stage of the questionnaire survey, probability sampling strategies were adopted. Stratified sampling was used for the identification of groups of construction firms with different financial qualification classes. The data sample was confined to construction firms with projects under construction during the period of the 5/12 Sichuan earthquake in 2008.

After identification of the sampling frame, a decision has to be made to identify the potential interviewees. With respect to this research, the construction team leader or project manager or senior manager was considered to be the most reliable source of knowledge about the companies and projects studied and what happened on site in the midst of the earthquake crisis. They were the key persons-in-charge who responded to the crisis with respect to the stakeholder communication process, hence they can be considered well placed to judge the effect of the variables/ factors that affected the communication performance during the crisis.

5.4.4 Questionnaire Data Collection

The use of questionnaires to collect data in most survey methods could be administrated in three ways: (1) self-administered postal questionnaire; (2) face-to-face interview; and (3) telephone interview (Robson 2002). For this research, a self-administered postal questionnaire was adopted as the main data collection method. It required the respondents to complete the predefined questionnaire and returned their replies through post.

In addition, face-to-face interviews using questionnaires were also conducted to ensure that all questions were answered and the information was accurate. The use of face-to-face interviews could also be advantageous in that the respondents could, if necessary, fully probed the meaning of the questions and have a chance to clarify any doubts with the researcher immediately. This approach also allowed general discussion and peripheral comments to be noted to add supporting contextual evidence (Walker 1997).

5.4.5 Questionnaire Data Analysis

A detailed analysis of the questionnaire returns was conducted using appropriate statistical analysis tools such as the Statistical Package for the Social Science (SPSS). Several statistical techniques were used to analyze the quantitative data and to interpret the results obtained from the questionnaire survey.

Based on the importance ratings on a five-point Likert scale given by the respondents, the mean importance ratings for each organizational variable were calculated. The formula for computing the mean importance rating is given below.

$$a_h = \frac{1(n_1) + 2(n_2) + 3(n_3) + 4(n_4) + 5(n_5)}{(n_1 + n_2 + n_3 + n_4 + n_5)} \tag{5.1}$$

where h is the attribute reference, a_h is the mean importance rating of the attribute h, and n_1, n_2, n_3, n_4, and n_5 are the number of respondents who indicated on the five-point Likert scale, the level of importance where 1–5 referred to "least important" to "most important" respectively.

Having calculated the mean importance ratings from the information provided by the respondent samples, the next step assessed how important the attributes were for the respondent population. One-tailed t test was utilized to test the significance of the variables.

In terms of bivariate statistical analysis, Pearson correlation coefficient was used in this research to analyze the possible relationship among the measurable organizational variables, and therefore provided some explanation or implication for the variables. Pearson correlation coefficient provided a measure of the linear association between two metric variables, denoted by r sample correlation (Tan 1995, 2002). If $r = 1$, there is perfect positive correlation; while if $r = -1$, there is perfect negative correlation.

In addition, variation analysis was applied to examine the variation of views on the organizational variables among construction firms according to their respective sizes. This was to evaluate the impact level of influencing factors/variables on the crisis response communication management under different organizational context.

5.5 Case-Study Development

In order to illustrate and further analyze the findings from the quantitative studies, case studies were conducted. There were several appropriate designs for case studies according to Yin (2003) and Winston (1997): exploratory, explanatory, and descriptive. In this research, the case-study approach was adopted to offer a more holistic and in-depth understanding of the crisis response process of the construction organizations and the ways in which their crisis managers and employees responded to a chaotic system and how they communicated about the risk, the threat, and predictability. For this type of explanatory research, the case-study approach was considered as an ideal approach (Feagin et al. 1991; Yin 2003) for the process of "explanation building", with considerable ability to generate answers to the question "why?" as well as the "what?" and "how?" questions (Robson 2002).

In the case-study phase, face-to-face interviews were conducted as the primary data collection technique. This is in-line with Robson's (2002) assertion that interviews enable researchers to find out what people know, what they do, and what they think or feel. Burawoy (1991) highlighted that the qualitative interview approach can reveal a dialectic interaction between interview findings and existing theories, and subsequently facilitate the reconstruction of a theory. In fact, it is common to find that the interview approach is one of the most preferred ways in collecting data for the development or refinement of a theoretical framework (Wang and Li 2007).

Documentation and archival records were used to serve as the secondary data collection tool. Documentation and archival records such as the crisis management plan, media accounts of the crisis, organizational records, annual reports, and financial reports provided a comprehensive source of information regarding the organizational communication management and overall crisis response performance that was valuable for the following data analysis process.

5.5.1 Case-Study Data Collection

Within the case studies, two construction firms in Chengdu city which were affected by the Sichuan earthquake crisis in 2008 were selected for investigation.

For case study 1, a state-owned construction firm with qualification of Class C1 was selected. This firm was one of the largest and most recognized local construction firms in Chengdu. With a total of 24 projects affected, its organizational communication structure encountered great challenges to ensure flexible information flow.

For case study 2, a private-owned construction firm with Class C2 qualification was selected. It represented a typical local construction firm in Chengdu which was privately owned and medium-sized. Based on the researcher's personal contacts and knowledge, this firm had developed crisis management plans covering a wide range of site scenarios, including earthquakes. In addition, it produced some level of flexibility and adaptivity in response to the earthquake, as a result of which it had been upgraded to Class C1 successfully in the few months following the earthquake crisis.

The most commonly used sources of evidence in doing case studies included documentation, archival records, interviews, direct observations, participant-observation, and physical artifacts (Yin 2003). In this research, the case-study data was collected mainly through the following sources: (a) a formal set of interviews with the project managers and/or crisis managers involved in the response to this crisis; (b) documentation including major newspapers and other popular media reports and the organization's crisis management plan; and (c) archival records, e.g., organizational internal records and financial reports.

The interviews were involved in-depth one-to-one discussions between the interviewees and the researcher and lasted around 2 h. The respondents were encouraged to recall the communication procedures and processes associated with the crisis response and to evaluate the organization's adaptive capability in response to the crisis from their own perspective. This was a quite common approach in complexity research (Hatch and Tsoukas 1997; Luhman and Boje 2001; Stacey 2001; Mitleton-Kelly 2005).

The interviews adopted a semi-structured approach. It was based on the use of an interview guide that comprised a list of predetermined questions. During the interview, the interviewer could skip questions or modify their order based on his/her discretion. Compared to structured interviews, which aimed at capturing precise data in order to explain behavior based on preestablished categories, unstructured interviews aimed to understand the complex behavior of members in an organization without limiting the field of inquiry (Fontana and Frey 1994). Thomas (1999) also found that "the danger of being in an interview with a list of questions written in stone is that the list becomes a crutch that hobbles the researcher in pursuing data" (p. 40). Whyte (1984) encouraged questions about specific events and the expansion on experiences with specific examples. Bearing the above point in mind, this research adopted semi-structured interviews, in that all respondents were asked a series of common questions (See Appendix C Interview Guide), but these were also open-ended. This semi-structured interview was adopted with the aim to foster flexibility to explore various areas and encourage the interviewees to recall the events associated with the crisis and evaluate the organizational variables from their own perspectives. This approach ensured some commonality across interviews while encouraging participants to expand on points they viewed as important, a condition which is difficult to achieve in both unstructured and structured interviews.

A review of the crisis management plan that was implemented during the crisis response was conducted to serve as an important data collection tool. Media accounts of the crisis and relevant reports were collected from the local and/or national press. A number of archival records such as organizational records, annual reports, and financial reports were also examined. All these information sources can help to create a more comprehensive understanding of the complex communication process and pattern during the crisis response stage.

5.5.2 Case-Study Data Analysis

For each case study, detailed written notes were taken during the interviews, which were transcribed and returned to the interviewees for review and correction. Audio-tape recording was also taken with the permission of interviewees and later fully transcribed.

The contents of the narrative by each interviewee were analyzed by content analysis, a research technique for the objective and systematic description of the manifest contents of communication (Berelson 1971). It involved categorizing the communication contents such as written notes or transcribed verbal narrative into its component parts and quantifying them, in order to derive meaning from them (Berg 2006). The analysis of each narrative enabled the researcher to probe decision making and communication processes experienced by the interviewees involved in the crisis response and to see how the interviewees assessed the influence of the organizational variables on their adaptive communication capability in the face of the crisis.

5.6 Summary

This research was conducted in four phases: (i) research model development; (ii) survey instrument development; (iii) case-study development; and (iv) research validation, which combined both the qualitative and quantitative approaches. This combination capitalized on the strengths and complements each approach, and thus provided a synergistic research design. Strategies have also been applied throughout the research process to reduce researcher's bias and influence on the findings. In the initial literature review phase, the conceptual framework has been discussed and consulted with scholars in the field of organizational management. Modifications and improvements have been made. In the following survey instrument phase, a pilot study has been conducted with some industry practitioners to ensure the completeness of the organizational variables in the survey questionnaire. In addition, various scientific statistical analysis techniques have been applied to the questionnaire survey data. Finally in the case study and validation phase, a set of face-to-face semi-structured interviews with project managers, crisis managers, and senior professionals have been conducted. All the respondents were asked to expand their professional views on the listed questions, based on their crisis experiences to ensure justification and validity of the results.

Chapter 6
Data Analysis

6.1 Introduction

This chapter provides detailed examination of the crisis response communication of Chengdu construction firms in the Sichuan earthquake in May 2008 in China. An in-depth analysis of the data collected from the survey, source documents, and interview sessions with professionals was structured according to the complexity-informed model. Based on the research design described in Chap. 5, a major survey of the Chengdu construction firms, preceded by a pilot study, was undertaken to ascertain which of the 59 organizational variables were important, and their level of importance.

After obtaining the responses from the survey, statistical analyses were undertaken to identify the important organizational variables. Two case studies relating to various issues of crisis response communication management in Chengdu construction firms were also presented.

6.2 The 2008 Sichuan Earthquake Crisis

A massive earthquake, which struck Sichuan province, a mountainous region in southwest China, on May 12 2008, thrusted China into the global spotlight. The earthquake hit Wenchuan County in Sichuan Province at 14:28 local time and was measured at 8.0 magnitude by the China Seismological Bureau and 7.9 magnitude by the US Geological Survey. Tremors from the earthquake were felt far and wide from Shanghai, Tianjin, Beijing cities, and some of the nearby countries in South East Asia. This earthquake was a large-magnitude event that caused unprecedented casualties and damage. Close to 90,000 people were classified either as fatalities or as unaccounted for. More than 4 million people were displaced and the number of collapsed or seriously damaged structures exceeded 25 million.

The epicenter was 73 km (45.6 miles) away from the urban Chengdu city, the capital of Sichuan Province. Urban Chengdu did not suffer dramatically from the

Z. Ying and L. Sui Pheng, *Project Communication Management in Complex Environments*,
DOI: 10.1007/978-981-4560-64-1_6,

earthquake in comparison with Wenchuan, Mianyang, Deyang, and other hard-hit regions. More heavily impacted areas within Chengdu were Dujiangyan and Pengzhou, which lie adjacent to Longmen Mountains. Seismologist experts testified that the geological structure of Chengdu Plain remains stable and that the earthquake and aftershocks that occurred along the Longmen mountain fault zone have not adversely affected urban Chengdu. The Chengdu Housing Administration has deemed that the earthquake's impact on Chengdu housing to be minimal, with collapsed structures and houses that were in danger of collapsing accounting for not more than 1 % of the total residences. However, there was no doubt that the Sichuan earthquake had unseated Chengdu's economy from its trend of steady growth.

6.3 Questionnaire Survey Results

To analyze and evaluate the significant variables, which influence the crisis response communication system within an organizational context, a fieldwork was conducted with Chengdu construction firms to understand their modes of operations.

6.3.1 Response Rate and Representativeness of Data

The questionnaire survey was restricted to the professionals who were involved in the crisis response management of on-going building projects in the event of the Sichuan earthquake. The selection process for the respondents was carried out based on the following criteria:

a. The professionals were restricted to those who were directly involved in communication management in response to the earthquake crisis.
b. The professionals were restricted to those from construction firms who were registered under the general building contractor category.
c. The construction firms were restricted to those who operated the building projects located in the metropolitan area of Chengdu City, which were affected by the 5/12 Sichuan earthquake.

In the fieldwork, 86 Chengdu construction firms were approached through email, telephone, and personal contacts. The survey packages that included the final questionnaire along with a covering letter stating the objectives of the study were sent to the intended persons in October 2009. A few rounds of personal discussion and face-to-face interviews with several respondents were also conducted. Responses were received between October 2009 and June 2010.

After checking over the completed questionnaires, 46 questionnaires were found suitable for data analysis. This yielded a response rate of around 53.6 %, which was considered to be at an acceptable level.

6.3.2 Data Processing

The first step in processing the data was to edit this data to ensure completeness, consistency, and readability. There was no missing information among the questionnaires and the ratings were also visually checked to ensure that no one respondent did not use one particular rating that might suggest that he or she had not thoughtfully assessed the variables. It was found that all the respondents had used a wide range on the five-point scale provided.

Following Tan (2002), after the responses were edited, the data were coded so that these could be further processed and categorized using a chosen software. The data received from the returned questionnaires were coded using the Statistical Package for the Social Sciences (SPSS). Various statistical techniques were applied to analyze the data.

6.4 Profile of Sample Firms and Respondents

The profiles of the sample firms and respondents were analyzed based on the useful responses, which provided a good representativeness of data for this study.

6.4.1 Characteristics of the Sample Construction Firms

The Chinese construction industry itself was undergoing a major evolution and transition. Construction enterprises in China were conventionally categorized into three groups: State-owned enterprises (SOEs), urban and rural collective-owned enterprises (COEs), and rural construction teams (RCTs) (Chen 1998; Low and Jiang 2003). The majority of the labor force in the construction sector during the 1980s and 1990s worked in SOEs and COEs. While many construction enterprises were small- and medium-sized enterprises (SMEs), the large-scale enterprises tended to be SOEs.

Since the mid 1990s, construction led the way when various reforms in China began in an effort to make the adjustment from a centrally planned operation to a market-oriented setting. A new classification system was implemented in 2002 with the aim to reduce the number of large-scale enterprises so that those with strong financial capability, sufficient technology base, talented executives, good management, and high productivity could come to the fore.

The new classification system categorized construction firms into three groups: general contractors, specialist contractors, and labor supply subcontractors, the latter representing a new category. It further divided general contractors into 12 different categories, specialist contractor into 60 types, and labor supply subcontractors into 13 forms. Within this classification system, the majority technology-focused construction enterprises would act more like general contractors and providers of project management services, while labor supply subcontractors and specialist contractors, such as those involved in interior fittings would become major providers of labors.

In this research, the targeted group of construction firms in Chengdu was selected under the general building contractor's category from the Sichuan Construction Bureau's General Contractor Registry. According to the Sichuan Construction Bureau, the general building contractors were registered under four financial qualification classes: SC (Special Class) contractors—financial limit of RMB300 million; C1 (Class One) contractors—financial limit of RMB50 million; C2 (Class Two) contractors—financial limit of RMB20 million; and C3 (Class Three) contractors—financial limit of RMB6 million (SCJST 2006). This research study grouped SC and C1 contractors into LSEs, and C2 and C3 contractors into SMEs.

As shown in Fig. 6.1, of the total 46 sample firms in this study, 58.7 % were from the general building contractors who were registered in the SC and C1 categories, and 41.3 % in the C2 and C3 categories. This indicated that a fair representativeness of both LSEs and SMEs in Chengdu had been covered in the survey.

A list of the projects and services typically provided by the sample construction firms was investigated in the fieldwork, and the results are shown in Table 6.1. This shows that a large portion of the projects undertaken by the sample firms was in residential construction (42.6 %) and civil engineering (37.6 %). The remaining

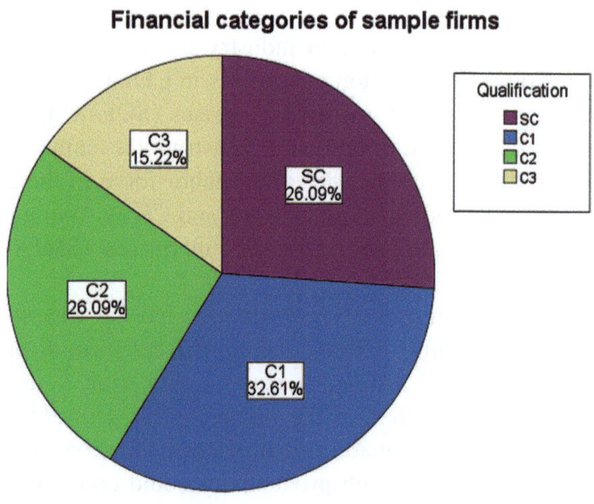

Fig. 6.1 Classifications of the sample firms

Table 6.1 Types of projects and services typically provided by the sample construction firms

Types of business and services provided	Frequency	Percentage (%)
Residential construction	43	42.6
Civil engineering	38	37.6
Commercial/office building	8	7.9
Industrial/factory construction	10	9.9
Others	2	2.0
Total	101	100.0

19.8 % of the firms were mainly involved with commercial and industrial construction. The extent of projects and services experiences might credit the respondents with the understanding and familiarity with various building project issues.

The survey results revealed, as shown in Table 6.2, that the majority (71.7 %) of the sample firms were privately owned enterprises, while 28.3 % of these were state-owned enterprises. This was acceptable because of the structural reforms in Chinese construction firms. Since the 1980s, the Chinese government has been encouraging the development of privately owned enterprises by taking effective measures to protect their legal rights and benefits. The state-owned enterprises were either directly under the administration of the Chinese central government, or under the administration of the Sichuan provincial government.

6.4.2 Characteristics of Respondents

All 46 respondents of the questionnaires were at the managerial or high professional levels in the Chengdu construction firms. A summary of the designation or position of the respondents is shown in Table 6.3. 80 % of the respondents were at top management or senior management position, while the other 20 % were either senior professional staffs or chief engineers. This means that the respondents were highly qualified in terms of seniority, and this contributed to the credibility of the survey results.

Table 6.4 shows that more than half of the respondents had been working in the construction industry for 10–20 years. In addition, the total average working experience of the respondents was 18 years. This indicates that the respondents had accumulated much working experience in the construction industry.

Table 6.2 Types of the sample construction firms

Firm type	Number	Percentage (%)
Private Ltd	33	71.7
State-owned	13	28.3
Others	0	0
Total	46	100.0

Table 6.3 Designation or position of respondents

Designation	Number	Percentage (%)
Top management (general manager, department director, production manager)	10	21.7
Senior management (senior manager, project manager, project executive manager, engineering project manager)	26	56.6
Middle management (senior professionals, chief engineers)	10	21.7
Total	46	100.0

Table 6.4 Years of practice of respondents

Years of practice	Number	Percentage (%)
1–10 years	5	10.9
10–20 years	24	52.2
>20 years	17	38.9
Total	46	100.0

The basic profiles of the respondents provided a fair representativeness of the data for this research study. The respondents were expected to be capable of articulating crisis response issues for the building projects in Chengdu in the aftermath of the earthquake, and the information they provided supported this study.

6.5 Project Characteristics

As mentioned earlier about the questionnaires, the survey was targeted at building projects in the metropolitan area of Chengdu city, which is around 80 km away from the earthquake epicenter. All 46 respondents indicated that the locations of their affected projects were inside the urban area.

6.5.1 Project Stage

The definition of "early, middle and end stages" of the building construction phase in this research was based on inputs from the industry practitioners involved in the pilot study. Generally, the early stage of the construction phase involves staff entry, site equipment deployment, and foundation completion. The middle stage includes the main structural construction works. And finally, the rest of the site works before delivery is considered as the final stage.

As shown in Table 6.5, about half of the building projects encountered the earthquake during the middle stage of the construction phase. The other 28.3 and 28.3 % of the projects were affected during the initial stage and end stage of the construction phase, respectively.

Table 6.5 Project stage

Project stage	Number	Percentage (%)
Early stage	13	28.3
Middle stage	20	43.5
End stage	13	28.3
Total	46	100.0

6.5.2 Project Contract Value

Among the 46 general contractor respondents, 38 of them provided information regarding the project contract value. The project value ranged between 1 and 300 million RMB.

As shown in Table 6.6, approximately 50 % of the building projects in the survey had contract value more than 50 million RMB. Moreover, 28.9 % of the projects had relatively smaller contract value less than 20 million RMB. Hence, building projects with a wide range of project contract values were studied in the survey, and a realistic representation of the crisis response issues for projects of different sizes could be ensured.

6.5.3 Perception of Critical Period in Response Crisis

In the survey questionnaire, the respondents were asked to provide their perceptions and experience regarding the crisis response critical period in the aftermath of the earthquake with respect to human safety and resource protection.

As shown in Table 6.7, the critical period in the aftermath of the earthquake for human safety was perceived by a vast majority of the respondents which was 97.8 %, as within the first hour. Only 2.2 % respondents indicated the critical period within the first day. It was found that human safety was the first priority in response to the earthquake for the Chengdu construction firms.

In terms of resource protection from pilfering and damages, the majority of the respondents which was 78.3 % perceived the response critical period as within the first day. The remaining 6.5 and 13.0 % of the respondents indicated the critical period as within the first hour and within first week, respectively. Only one respondent considered this to be within the first month.

Table 6.6 Project contract value

Project contract value	Number	Percentage (%)
<20 (Million RMB)	11	28.9
20–50 (Million RMB)	8	21.1
>50 (Million RMB)	19	50
Total	38	100.0

Table 6.7 Critical period in the aftermath of the earthquake

Critical period	Safety		Resource protection	
	Number	Percent (%)	Number	Percent (%)
Within the first hour	45	97.8	3	6.5
Within the first day	1	2.2	36	78.3
Within the first week	0	0.0	6	13.0
Within the first month	0	0.0	1	2.2
Total	46	100.0	46	100.0

6.6 Organizational Variables Affecting Crisis Communication Management in Chengdu Construction Firms

Following the complexity-informed framework, organizational variables which affected communication management in response to the earthquake in Chengdu construction firms were analyzed in details in this section. In the first step, organizational variables which influenced the level of different complex control parameters were analyzed, respectively, and correlations between the variables which affected the same control parameters were discussed; in the second step, variations of the views on all the 59 organizational variables among different size/scale of Chengdu construction firms (i.e., LSE vs. SME) were examined.

6.6.1 Organizational Variables of CP1

Information flow pattern within an organizational communication network influenced the crisis response efficiency and adaptability of the construction firms. There were various organizational variables which affected the information flow and they included organizational structure, organizational culture, information technology management, leadership style, sensing-making capability, and communication skill capability of the organization and its members.

To assess how important the organizational variables were, statistical tests of the mean were carried out to check whether the population would consider the variables to be important or otherwise. From the table of critical values of t-distribution, for degrees of freedom = 45 (46−1), the level of significance for a two-tailed test at 0.05, the t value is 2.014. This means that if the calculated t value in Table 6.8 was larger than 2.014, the statistical null hypothesis (H_0:$\mu \leq 3$, where μ is *thepopulationmean*) that the variable was unimportant was rejected and the alternative hypothesis (H_1:$\mu > 3$) was accepted. In other words, it could be concluded that the variable was important.

As shown in Table 6.8, the most significant variables which affected the CP1 were identified, in the order of importance, as follows:

Table 6.8 Organizational variables of CP1: rate of information flow

Organizational variables	Mean	Std. deviation	t value	95 % confidence interval of the difference	
				Lower	Upper
OV1.1 A flat hierarchy structure	4.6304	0.48802	22.659	1.4855	1.7754
OV1.2 A highly bureaucracy top-down structure	3.5217	1.24256	2.848	0.1527	0.8907
OV1.3 An ad hoc decentralized structure	3.7609	0.60313	8.556	0.5818	0.9400
OV1.4 A routine administrative structure	2.1304	0.74859	−7.878	−1.0919	−0.6473
OV2.1 Learning environment encouraging constantly to acquire and communicate knowledge and share critical crisis information	4.5435	0.50361	20.787	1.3939	1.6930
OV2.2 Collaborative culture fostering interaction and relationship between project participants	4.2174	0.62939	13.119	1.0305	1.4043
OV3.1 Usage of intranet within the company with e-mail communication system	4.4130	0.65238	14.690	1.2193	1.6068
OV3.2 A dedicated crisis or emergency website in the company's intranet system	3.6739	0.92025	4.967	0.4006	0.9472
OV3.3 Usage of proper modern information technology to facilitate communication	4.4565	0.62206	15.881	1.2718	1.6413
OV3.4 Emergency phone numbers and the site location are posted and clearly marked beside all site phones	4.0435	0.63093	11.217	0.8561	1.2308
OV3.5 An innovative integration information system is adopted	4.3696	0.67852	13.690	1.1681	1.5711
OV3.6 Employees have proper instructing information they need to know to protect themselves physically	4.4783	0.50505	19.852	1.3283	1.6282
OV3.7 Feedback and response to the issues are prompt in a timely fashion	4.4739	0.47396	23.954	1.5332	1.8147
OV4.1 A flexible communication management process and procedures to allow for open communication	4.4087	0.49344	22.112	1.4622	1.7552
OV4.2 A rigid and centralized communication management process and procedures	3.1739	1.08124	1.091	−0.1472	0.4950
OV4.3 Contractual rules and procedures are strictly followed during the response stage	3.1087	0.64043	1.151	−0.0815	0.2989
OV4.4 Project managers or leaders are involved in great details in decision making	2.7826	0.72765	−2.026	−0.4335	−0.0013

(continued)

Table 6.8 (continued)

Organizational variables	Mean	Std. deviation	t value	95 % confidence interval of the difference	
				Lower	Upper
OV4.6 A decentralized decision-making process	3.6087	0.53658	7.694	0.4493	0.7680
OV4.7 A delegation of decision authority style (empowerment), employees have amount of autonomy to make on-the-spot decisions during the crisis	4.6087	0.49344	22.112	1.4622	1.7552
OV4.8 A proactive decision-making process	3.5435	0.62206	5.926	0.3587	0.7282
OV4.12 A control of authority leadership style (centralized)	3.1087	0.82269	0.896	−0.1356	0.3530
OV4.13 A risk-taking preference leadership style (over-confident)	1.8913	0.52613	−14.292	−1.2649	−0.9525
OV4.14 A habit-path resort preference leadership style (empiricism)	3.0217	0.77428	0.190	−0.2082	0.2517
OV4.15 A long-term decision-making vision	3.8913	0.48204	12.541	0.7482	1.0345
OV5.1 Project participants feel a high level of trust when communicating openly and truthfully	4.3913	0.68242	13.828	1.1886	1.5940
OV5.13 Our employees maintain a good interpersonal working relationship	3.8478	0.75916	7.574	0.6224	1.0733
OV6.1 Our employees possess good communication skills to discourse and convey their opinions	4.1304	0.58152	13.184	0.9577	1.3031

- OV1.1: A flat hierarchy structure;
- OV4.7: A delegation of decision authority style;
- OV2.1: Learning environment encouraging constantly to acquire and communicate knowledge and share critical crisis information;
- OV3.6: Employees have proper instructing information they need to know to protect themselves physically;
- OV3.7: Feedback and response to the issues are prompt in a timely fashion;
- OV3.3: Usage of proper modern information technology to facilitate communication;
- OV3.1: Usage of intranet within the company with e-mail communication system;
- OV4.1: A flexible communication management process and procedures to allow for open communication;
- OV5.1: Project participants feel a high level of trust when communicating openly and truthful;
- OV3.5: An innovative integration information system is adopted;
- OV2.2: Collaborative culture fostering interaction and relationship between project participants;
- OV6.1: Our employees possess good communication skills to discourse and convey their opinions;
- OV3.4: Emergency phone numbers and the site location are posted and clearly marked beside all site phones;
- OV4.15: A long-term decision-making vision;
- OV5.13: Our employees maintain a good inter-personal working relationship;
- OV1.3: An ad hoc decentralized structure;
- OV3.2: A dedicated crisis or emergency website in the company's intranet system;
- OV4.6: A decentralized decision-making process;
- OV4.8: A proactive decision-making process; and
- OV1.2: A highly bureaucracy top-down structure.

The remaining seven variables were recognized as of no significant influence to the crisis communication management:

- OV4.14 A habit-path resort preference leadership style;
- OV4.13 A risk-taking preference leadership style;
- OV4.12 A control of authority leadership style;
- OV4.4 Project managers or leaders are involved in great details in decision making;
- OV4.3 Contractual rules and procedures are strictly followed during the response stage;
- OV4.2 A rigid and centralized communication management process and procedures; and
- OV1.4 A routine administrative structure.

To evaluate the possible relationship between the significant organizational variables which affected CP1, Pearson's correlation coefficients of the pair variables were calculated and the statistical significance was used to distinguish the possible relationship.

From the correlation analysis in Table 6.9, the close relationship between OV1.1 and OV4.1 indicates that a flat hierarchy organizational structure might influence the flexibility of the organization's communication management process, and as a result, allowed for an open communication environment within the organization.

The correlation between OV2.2 and OV4.7 might possibly imply that a delegation of authority control leadership style affected the organizational culture of fostering collaboration and interaction between project members. The correlation between OV2.2 and OV3.5 further implied that the deployment of an integration information system also links to the organizational collaborative culture. Therefore, the analysis suggests that development of a collaborative culture for effective project communication in response to a crisis required the firm to enhance its information technology management and the crisis manager's capability to control a certain degree of authority.

The relatively close relationship between OV5.1 and other different organizational variables including OV1.1, OV2.1, OV4.1, and OV5.13 also suggests that the firm's organizational structure, a knowledge sharing culture and flexible communication management were highly related to its employees' capability of trustworthy and trusty interaction under a crisis situation. In return, a high level of trust facilitated crisis communication between the project participants to implement and transfer the information about their crisis activities.

The correlation analysis in Table 6.9 also shows a significant relationship between OV3.1, OV3.2, and OV3.5. The development of information technology could greatly increase the exchange rate of the information in the crisis, which was essential to shorten the feedback loops between project members. It was understandable that the usage of the intranet, availability of dedicated crisis website, and adoption of integration information system were closely related to each other for improved crisis information processing capability of the construction firm.

The correlation analysis in Table 6.9 reflected a number of negative correlations, one of which was significant—the significant negative correlation between OV1.1 and OV1.2 confirmed the recognized relation that a flat hierarchical structure means less of a highly bureaucratic top-down structure (Robbins and Judge 2007).

6.6.2 *Organizational Variables of CP2*

Connectivity, degree, and quality of the connections were critical parameters that influenced a complex adaptive system's ability to self-organize. In terms of the crisis response communication system for the construction firms, various

Table 6.9 Correlation analysis of organizational variables of CP1

	OV1.1	OV1.2	OV1.3	OV2.1	OV2.2	OV3.1	OV3.2	OV3.3	OV3.4	OV3.5	OV3.6	OV3.7	OV4.1	OV4.6	OV4.7	OV4.8	OV4.15	OV5.1	OV5.13	OV6.1
OV1.1	–	−0.408**	−0.107	0.202	0.140	0.211	0.069	−0.018	0.053	0.287	−0.159	0.236	0.560**	−0.110	0.217	0.018	0.014	0.310*	−0.135	0.130
OV1.2		–	0.022	0.212	−0.120	−0.190	−0.106	0.145	−0.086	−0.156	0.008	−0.082	0.240	−0.054	−0.095	−0.116	−0.126	−0.199	0.081	−0.219
OV1.3			–	−0.167	−0.070	−0.026	−0.024	0.120	0.203	−0.051	0.019	0.032	−0.097	0.048	−0.023	−0.120	0.061	−0.146	0.161	−0.226
OV2.1				–	−0.030	0.181	0.103	−0.029	−0.006	−0.081	0.206	−0.079	0.149	−0.018	−0.109	0.171	0.066	0.684**	0.105	0.132
OV2.2					–	0.209	0.248	0.252	−0.004	0.688**	−0.064	−0.130	0.137	−0.137	0.665*	−0.081	0.006	0.160	−0.069	0.042
OV3.1						–	0.599**	0.101	−0.207	0.300*	0.062	−0.130	0.030	−0.036	0.237	−0.073	0.146	0.078	−0.019	0.265
OV3.2							–	0.072	0.178	0.624**	0.008	0.107	−0.091	−0.039	0.153	0.045	0.269*	0.120	−0.254	0.164
OV3.3								–	0.005	0.013	0.138	0.139	0.161	−0.252	0.161	−0.183	0.095	0.146	0.197	−0.107
OV3.4									–	0.125	0.073	0.194	−0.016	0.051	0.070	0.052	0.181	−0.092	−0.079	−0.076
OV3.5										–	−0.073	0.176	−0.155	−0.143	−0.023	−0.013	0.193	0.209	−0.036	−0.012
OV3.6											–	0.016	0.189	−0.196	0.054	0.003	0.036	0.219	0.126	0.161
OV3.7												–	0.107	0.011	0.202	0.011	0.133	−0.009	−0.141	−0.004
OV4.1													–	0.255	0.179	−0.088	0.004	0.733**	0.090	0.137
OV4.6														–	0.048	0.218	0.176	−0.240	−0.040	0.025
OV4.7															–	0.057	0.097	0.067	0.075	0.014
OV4.8																–	0.201	−0.093	−0.197	0.107
OV4.15																	–	−0.206	−0.046	0.052
OV5.1																		–	0.354*	0.173
OV5.13																			–	−0.055
OV6.1																				–

** Correlation is significant at the 0.01 level (1-tailed)
* Correlation is significant at the 0.05 level (1-tailed)

organizational variables could be perceived to affect the level and richness of the connections and interdependencies between the project members and participants. In this section, 14 organizational variables were analyzed with the fieldwork results presented in Table 6.10.

Based on the t-test results, Table 6.10 shows the most significant variables which affected CP2, in the order of importance, as follows:

- OV1.1: A flat hierarchy structure;
- OV5.5: Project participants demonstrate transparency of intent and straightforward actions;
- OV2.1: Learning environment encouraging constantly to acquire and communicate knowledge and share critical crisis information;
- OV5.3: Project participants show respects to one another, and hold accountable of each other;
- OV4.1: A flexible communication management process and procedures to allow for open communication;
- OV5.1: Project participants feel a high level of trust when communicating openly and truthfully;
- OV2.2: Collaborative culture fostering interaction and relationship between project participants;
- OV6.1: Our employees possess good communication skills to discourse and convey their opinions;
- OV5.7: Problems can be recognized at the very first sign and dealt with quickly, without transfer of blame to those who are not responsible; and
- OV1.2: A highly bureaucracy top-down structure.

The remaining four organizational variables were perceived to be of no significant influence:

- OV4.2 A rigid and centralized communication management process and procedures;
- OV1.4 A routine administrative structure;
- OV5.2 Project participants offer help and provide reciprocal service; and
- OV5.6 Project participants have a very strong feeling of obligation and friendship.

It was found that the most significant variable affecting CP2 was OV1.1, which was of the same order of importance to CP1. This indicates the interdependencies of CP1 and CP2 parameters in that the rate of information flow itself could be dictated by connectivity within the system. A flat hierarchy structure was evidently related to the connectivity and interactivity between project members, and therefore to the rate of the information flow among the project members. The more hierarchical the organizational structure, lower level of connectivity and interactivity might occur between organizational members and hence, the slower the rate of information flow might be.

The correlation analysis in Table 6.11 indicates that the culture of collaboration and interaction in a construction firm influenced the capability of mutual respect

Table 6.10 Organizational variables of CP2: degree of connections and interdependencies

Organizational variables	Mean	Std. deviation	t value	95 % confidence interval of the difference	
				Lower	Upper
OV1.1 A flat hierarchy structure	4.6304	0.48802	22.659	1.4855	1.7754
OV1.2 A highly bureaucracy top-down structure	3.5217	1.24256	2.848	0.1527	0.8907
OV1.4 A routine administrative structure	2.1304	0.74859	−7.878	−1.0919	−0.6473
OV2.1 Learning environment encouraging constantly to acquire and communicate knowledge and share critical crisis information	4.5435	0.50361	20.787	1.3939	1.6930
OV2.2 Collaborative culture fostering interaction and relationship between project participants	4.2174	0.62939	13.119	1.0305	1.4043
OV4.1 A flexible communication management process and procedures to allow for open communication	4.4087	0.49344	22.112	1.4622	1.7552
OV4.2 A rigid and centralized communication management process and procedures	3.1739	1.08124	1.091	−0.1472	0.4950
OV5.1 Project participants feel a high level of trust when communicating openly and truthfully	4.3913	0.68242	13.828	1.1886	1.5940
OV5.2 Project participants offer help and provide reciprocal service	3.1087	0.84927	0.868	−0.1435	0.3609
OV5.3 Project participants show respects to one another, and hold accountable of each other	4.4783	0.50505	19.852	1.3283	1.6282
OV5.5 Project participants demonstrate transparency of intent and straightforward actions	4.5352	0.58525	17.887	1.3697	1.7173
OV5.6 Project participants have a very strong feeling of obligation and friendship	2.0000	0.69921	−9.700	−1.2076	−0.7924
OV5.7 Problems can be recognized at the very first sign and dealt with quickly, without transfer of blame to those who are not responsible	4.0217	0.44667	15.514	0.8891	1.1544
OV6.1 Our employees possess good communication skills to discourse and convey their opinions	4.1304	0.58152	13.184	0.9577	1.3031

Table 6.11 Correlation analysis of organizational variables of CP2

	OV1.1	OV1.2	OV2.1	OV2.2	OV4.1	OV5.1	OV5.3	OV5.5	OV5.7	OV6.1
OV1.1	–	−0.408**	0.202	0.140	−0.060	0.310*	0.212	0.108	−0.064	0.130
OV1.2		–	0.212	−0.120	0.240	−0.199	−0.194	−0.215	−0.101	−0.219
OV2.1			–	−0.030	0.149	0.014	0.266*	0.031	0.243	0.132
OV2.2				–	0.137	0.160	0.725**	0.655**	−0.175	0.042
OV4.1					–	0.733**	0.533**	0.214	0.341*	0.137
OV5.1						–	0.212	0.202	0.190	0.373**
OV5.3							–	0.078	0.051	0.237
OV5.5								–	−0.046	0.375**
OV5.7									–	0.331*
OV6.1										–

** Correlation is significant at the 0.01 level (1-tailed)
* Correlation is significant at the 0.05 level (1-tailed)

and trust, transparency of intent, and straightforward actions between project members, as implied by the relatively close relationship between OV2.2, OV5.3, and OV5.5. These organizational variables largely affected the communication effectiveness in a crisis response stage and should be well informed by the project managers.

The correlation between OV4.1 and other variables including OV5.1, OV5.3, and OV5.7 implied that the flexible communication management process was likely to be related to the capability of organizational members to communicate with each other in the crisis situation based on the principle of high level of trust, respect, and the low level of making someone a scapegoat.

It is also shown in Table 6.11 that there was close relationship between OV6.1 with the variables of OV5.1, OV5.5, and OV5.7. This further indicated that the employee's communication skills to discourse and convey their opinion were linked to their communicative relationship in terms of the level of trust, transparency of intent, and transfer of blame.

6.6.3 Organizational Variables of CP3

Level of diversity among the members within the crisis response communication system was critical for a smooth communication operation. Eleven organizational variables affecting the level of diversity control parameter were identified and taken into the analysis in the fieldwork, with the results presented in Table 6.12.

The statistical results of two-tailed t-test showed that the following organizational variables were regarded as important variables affecting CP3 in the order of their mean importance:

- OV2.6 Develop crisis goal alignment;
- OV2.5 Employees are fully aware of the emergency communication procedures and have crisis-prepared mindset;
- OV3.5 An innovative integration information system is adopted;
- OV2.4 Understand and commit to a shared vision;
- OV2.7 Facilitate a climate of constant change, conflict, and diversity;
- OV2.8 Encourage continuous discourse of different points of view;
- OV6.1 Our employees possess good communication skills to discourse and convey their opinions;
- OV5.8 Our employees are fully aware of their risks and obligations as well as others;
- OV4.16 Constructive conflict management for a well-managed conflict; and
- OV6.2 Our employees are capable of coping and resolving differences in opinions.

Table 6.12 Organizational variables of CP3: levels of diversity among agents of system

Organizational variables	Mean	Std. deviation	t value	95 % confidence interval of the difference	
				Lower	Upper
OV2.4 Understand and commit to a shared vision	4.1304	0.65349	11.732	0.9364	1.3245
OV2.5 Employees are fully aware of the emergency communication procedures and have crisis-prepared mindset	4.4130	0.68560	13.979	1.2094	1.6166
OV2.6 Develop crisis goal alignment	4.6087	0.49344	22.112	1.4622	1.7552
OV2.7 Facilitate a climate of constant change, conflict, and diversity	3.6622	0.56534	7.469	−0.1345	0.4388
OV2.8 Encourage continuous discourse of different points of view	3.6087	0.59522	6.927	−0.1275	0.3448
OV3.5 An innovative integration information system is adopted	4.3696	0.67852	13.690	1.1681	1.5711
OV4.16 Constructive conflict management for a well-managed conflict	3.8478	0.36316	15.834	0.7400	0.9557
OV4.17 A conflict avoidance and elimination preference	3.0217	0.25726	0.573	−0.0547	0.0981
OV5.8 Our employees are fully aware of their risks and obligations as well as others	3.8913	0.60473	9.996	0.7117	1.0709
OV6.1 Our employees possess good communication skills to discourse and convey their opinions	4.1304	0.58152	13.184	0.9577	1.3031
OV6.2 Our employees are capable of coping and resolving differences in opinions	3.5652	0.50121	7.649	0.4164	0.7141

According to the views of the respondents, it was not statically significant or less important for the other three variables in the crisis response communication management to affect CP3:

- OV4.17 A conflict avoidance and elimination preference.

The correlation analysis as shown in Table 6.13 indicates a number of relationships between the organizational variables. Organizational variable OV5.8 was linked to OV2.4 and OV2.6, while OV2.4, OV2.6, and OV2.5 were linked to each other. This possibly implied that employees' capability and awareness of their own risks and obligations as well as others in a crisis situation brought certain influence to the development of a common crisis goal and a shared vision and organizational culture. This could be beneficial for the professionals to learn from each other's behavior and adapt to the behavior of others.

It was also suggested that the alignment of common crisis goals, development of a shared vision, as well as explicit and implicit crisis emergency communication procedures linked together to influence crisis communication effectiveness to some extent.

Table 6.13 Correlation analysis of organizational variables of CP3

	OV2.4	OV2.5	OV2.6	OV2.7	OV2.8	OV3.5	OV4.16	OV4.17	OV5.8	OV6.1	OV6.2
OV2.4	–	0.326*	0.437**	−0.067	0.143	0.139	0.085	−0.017	0.505**	−0.046	0.313*
OV2.5		–	0.229*	0.172	−0.125	0.190	0.158	0.200	0.108	0.052	0.146
OV2.6			–	−0.059	0.094	−0.023	0.032	0.069	0.676**	0.104	0.105
OV2.7				–	−0.109	0.289*	−0.123	0.165	−0.123	0.081	0.048
OV2.8					–	−0.158	−0.018	0.097	0.056	0.065	−0.102
OV3.5						–	0.537**	0.208	−0.062	−0.012	−0.040
OV4.16							–	−0.036	0.125	0.107	−0.005
OV4.17								–	−0.127	−0.019	0.075
OV5.8									–	0.231	−0.086
OV6.1										–	0.199
OV6.2											–

** Correlation is significant at the 0.01 level (1-tailed)
* Correlation is significant at the 0.05 level (1-tailed)

The close relationship between OV3.5 and OV4.16 indicated that implementation of an integration knowledge information system was related to conflict management between professionals. This further implied that to facilitate constructive conflict management, construction firms should consider the adoption of an innovative information system for effective information sharing and communication between project participants with different professional backgrounds.

6.6.4 Organizational Variables of CP4

Fifteen organizational variables which could influence the level of contained anxiety and tension in response to a crisis event were taken into consideration. The fieldwork results are presented in Table 6.14.

Based on the statistical results of the two-tailed t test, the following organizational variables were identified as significant variables of CP4 in the order of their mean importance:

- OV4.11 An influencing, motivating, and inspiring leadership style;
- OV4.18 Encourage a defined sense of purpose;
- OV5.10 Our employees tend to self-reflect about the crisis situation and are willing to engage in double-loop learning;
- OV5.4 Project participants are confident and show a positive attitude to the project organization;
- OV2.3 Open-mindedness climate encourages mutually trusting and honest interaction;
- OV3.5 An innovative integration information system is adopted;
- OV6.5 Our employees possess functional knowledge to deal with paradox and ambiguities inherent in crisis situations;
- OV6.4 Our employees possess capability to cope with pressure inherent in crisis situations;
- OV5.13 Our employees maintain a good interpersonal working relationship;
- OV5.11 Our employees are willing to raise issues and take risks;
- OV6.3 Our employees possess good problem-solving and analysis skills to handle unforeseen circumstances;
- OV5.12 Our employees are willing to accept constructive feedback; and
- OV4.19 Crisis management/communication plan (if applicable) is strictly followed and given credit to.

The statistical results identified the remaining two organizational variables with less importance and not significant:

- OV4.12 A control of authority leadership style (centralized); and
- OV5.9 Develop defensive mechanism strategies to avoid anxiety.

Table 6.14 Organizational variables of CP4: contained anxiety and tension

Organizational variables	Mean	Std. deviation	t value	95 % confidence interval of the difference	
				Lower	Upper
OV2.3 Open-mindedness climate encourages mutually trusting and honest interaction	4.3696	0.74113	12.533	1.1495	1.5897
OV3.5 An innovative integration information system is adopted	4.3696	0.67852	13.690	1.1681	1.5711
OV4.11 An influencing, motivating, and inspiring leadership style	4.5652	0.54374	19.524	1.4037	1.7267
OV4.12 A control of authority leadership style (centralized)	3.1087	0.82269	0.896	−0.1356	0.3530
OV4.18 Encourage a defined sense of purpose	4.4565	0.62206	15.881	1.2718	1.6413
OV4.19 Crisis management/communication plan (if applicable) is strictly followed and given credit to	3.4783	0.50505	6.423	0.3283	0.6282
OV5.4 Project participants are confident and show a positive attitude to the project organization	4.4870	0.49782	21.621	1.4391	1.7348
OV5.9 Develop defensive mechanism strategies to avoid anxiety	3.0217	0.57693	0.256	−0.1496	0.1931
OV5.10 Our employees tend to self-reflect about the crisis situation and are willing to engage in double-loop learning	4.4870	0.49782	21.621	1.4391	1.7348
OV5.11 Our employees are willing to raise issues and take risks	3.8261	0.48554	11.539	0.6819	0.9703
OV5.12 Our employees are willing to accept constructive feedback	3.5435	0.50361	7.319	0.3939	0.6930
OV5.13 Our employees maintain a good inter-personal working relationship	3.8478	0.75916	7.574	0.6224	1.0733
OV6.3 Our employees possess good problem-solving and analysis skills to handle unforeseen circumstances	3.7391	0.49147	10.200	0.5932	0.8851
OV6.4 Our employees possess capability to cope with pressure inherent in crisis situations	4.0435	0.46935	15.079	0.9041	1.1829
OV6.5 Our employees possess functional knowledge to deal with paradox and ambiguities inherent in crisis situations	4.0652	0.38885	18.579	0.9497	1.1807

The correlation analysis of the significant organizational variables affecting CP4 is shown in Table 6.15. As shown by the relatively close relationship between OV2.3 and variables that included OV5.4, OV5.10, and OV5.11, an open-minded climate encouraged trusting interaction. This relates to the organization and its member's sense-making capability in terms of positive attitude, willingness to take risks, and self-reflection in response to a crisis. Hence, a trusting environment might harness the members of an organization to contain their level of anxiety by enhancing their sense-making capabilities.

It was also found that there existed a close relationship between OV2.3 and OV4.18. This implied that the leadership of encouraging the sense of purpose influenced the development of an open-minded and worthy fostering culture within an organization.

In addition, correlations between OV4.11 and the variables of OV6.4 and OV6.5 suggested that an influencing and motivating style of leadership was linked to the employee's skills capability in terms of coping with pressure, paradox, and ambiguities in crisis situations. This implied that when dealing with the anxiety and stress for project members, the project leader might consider showing tolerance for paradox and to influencing project members in response to an adaptive challenge.

6.6.5 Organizational Variables of CP5

Power distribution, when exercised in the crisis response communication system in an organization, greatly influenced the organization's ability to move to the edge of chaos. Different organizational variables might affect power distribution and the level of power differentials within an organization. In this section, 14 major organizational variables which affected CP5 were analyzed, and their degrees of importance examined. The fieldwork results are presented in Table 6.16.

The statistical results of the two-tailed t test in Table 6.16 showed that the following organizational variables were regarded as significant variables of CP5 in the order of their mean importance:

- OV1.1 A flat hierarchy structure;
- OV4.7 A delegation of decision authority style;
- OV2.1 Learning environment encouraging constantly to acquire and communicate knowledge and share critical crisis information;
- OV5.3 Project participants show respect to one another, and are accountable to each other;
- OV4.21 Manager's ability to Influence over others;
- OV2.3 Open-mindedness climate encourage mutually trusting and honest interaction;
- OV4.20 Manager's expertise to deal with crisis contingency;
- OV1.3 An ad hoc decentralized structure;

Table 6.15 Correlation analysis of organizational variables of CP4

	OV2.3	OV3.5	OV4.11	OV4.18	OV4.19	OV5.4	OV5.10	OV5.11	OV5.13	OV6.3	OV6.4	OV6.5
OV2.3	–	−0.078	0.242	0.349**	−0.067	0.542**	0.322*	0.291*	0.221	0.210	−0.175	−0.240
OV3.5		–	−0.217	0.013	−0.068	0.130	0.196	−0.070	0.036	−0.038	0.137	0.244
OV4.11			–	0.168	0.208	0.225	−0.021	0.212	0.267*	0.108	0.498**	0.332*
OV4.18				–	0.068	0.164	0.120	0.069	0.124	−0.011	0.146	0.034
OV4.19					–	0.184	0.096	0.016	0.152	0.156	−0.090	0.177
OV5.4						–	0.152	0.156	0.159	0.076	0.102	0.142
OV5.10							–	−0.120	0.100	0.186	0.102	0.027
OV5.11								–	0.013	0.101	−0.064	0.092
OV5.13									–	0.010	−0.018	−0.116
OV6.3										–	−0.142	0.091
OV6.4											–	0.106
OV6.5												–

** Correlation is significant at the 0.01 level (1-tailed)
* Correlation is significant at the 0.05 level (1-tailed)

Table 6.16 Organizational variables of CP5: Power differentials

Organizational variables	Mean	Std. deviation	t value	95 % confidence interval of the difference	
				Lower	Upper
OV1.1 A flat hierarchy structure	4.6304	0.48802	22.659	1.4855	1.7754
OV1.2 A highly bureaucratic top-down structure	3.5217	1.24256	2.848	0.1527	0.8907
OV1.3 An ad hoc decentralized structure	3.7609	0.60313	8.556	0.5818	0.9400
OV2.1 Learning environment encouraging constantly to acquire and communicate knowledge and share critical crisis information	4.5435	0.50361	20.787	1.3939	1.6930
OV2.3 Open-mindedness climate encourage mutually trusting and honest interaction	4.3696	0.74113	12.533	1.1495	1.5897
OV4.5 A participative decision-making process (democratic)	2.9565	0.91788	−0.321	−0.3161	0.2291
OV4.7 A delegation of decision authority style (empowerment), employees have autonomy to make on-the-spot decisions during the crisis	4.6087	0.49344	22.112	1.4622	1.7552
OV4.9 Most decisions are made through consensus	2.4565	0.78050	−4.723	−0.7753	−0.3117
OV4.10 The critical decisions are made by some dominant elites who have the power to delegate decisions	3.5000	0.72265	4.693	0.2854	0.7146
OV4.12 A control of authority leadership style (centralized)	3.1087	0.82269	0.896	−0.1356	0.3530
OV4.20 Manager's expertise to deal with crisis contingency	4.2174	0.51264	16.106	1.0652	1.3696
OV4.21 Manager's ability to Influence over others	4.3696	0.64494	14.403	1.1780	1.5611
OV5.3 Project participants show respect to one another, and are accountable to each other	4.4783	0.50505	19.852	1.3283	1.6282
OV6.6 Some of our employees possess competent technical skills and expertise that would be difficult to replace	3.4130	0.58027	4.828	0.2407	0.5854

- OV4.10 The critical decisions are made by some dominant elites who have the power to delegate decisions;
- OV1.2 A highly bureaucratic top-down structure; and
- OV6.6 Some of our employees possess competent technical skills and expertise that would be difficult to replace.

The remaining three organizational variables were regarded as less important:

- OV4.5 A participative decision-making process;
- OV4.9 Most decision are made through consensus; and
- OV4.12 A control of authority leadership style.

The correlation analysis as shown in Table 6.17 indicates the close relationship between OV4.7, OV4.20, and OV4.21. Project manager's ability to deal with contingency and uncertainty, the ability to influence others, and a delegation of authority leadership style were important power characteristics that project managers or top managers could exercise during the project life cycle. Therefore, these variables could be considered together when the project managers developed crisis communication strategies.

It was also found in the analysis that the variables OV1.2 and OV4.10 were significantly correlated. This implied that a bureaucratic top-down organizational structure was related to management strategy with an emphasis on quickly regaining control by the power elite/spokeman in response to different crisis issues.

6.6.6 All Organizational Variables

In this section, all the 59 organizational variables were tabulated according to their means and standard deviations, as shown in Table 6.18.

With the level of significance for a two-tailed test at 5 %, it was observed from Table 6.18 that there were ten organizational variables (OV2.7, OV2.8, OV4.2, OV4.3, OV4.5, OV4.12, OV4.14, OV4.17, OV5.2, OV5.9), where the differences between the population means and the hypothesized means were considered not to be statistically significant. Therefore, the null hypothesis that the variable was unimportant was accepted for those ten organizational variables.

Furthermore, the orders of the organizational variables in terms of the mean importance were extracted and categorized into the five most important ones as shown in Table 6.19. The results suggested that a flat hierarchy structure, a delegation of decision authority leadership style, development of crisis goal alignment culture, an influencing leadership style, a learning organizational culture, and transparency of intent and straightforward actions were considered as the most important organizational variables perceived by the respondents.

The most significant organizational variables are discussed in Sects. 6.6.6.1–6.6.6.6.

Table 6.17 Correlation analysis of organizational variables of CP5

	OV1.1	OV1.2	OV1.3	OV2.1	OV2.3	OV4.7	OV4.10	OV4.20	OV4.21	OV5.3	OV6.6
OV1.1	–	−0.408**	−0.107	0.202	0.140	0.217	−0.110	−0.027	0.091	0.212	0.159
OV1.2		–	0.022	0.212	0.244	−0.095	0.544**	0.111	0.170	−0.194	−0.036
OV1.3			–	−0.167	−0.045	−0.023	0.180	−0.116	−0.025	−0.027	−0.029
OV2.1				–	0.224	−0.109	0.092	0.137	0.217	0.266*	0.051
OV2.3					–	−0.021	−0.021	0.076	0.173	0.111	0.023
OV4.7						–	0.000	0.268	0.295*	−0.035	0.189
OV4.10							–	−0.060	−0.167	−0.244	−0.085
OV4.20								–	0.591**	0.104	−0.010
OV4.21									–	0.128	−0.001
OV5.3										–	0.008
OV6.6											–

** Correlation is significant at the 0.01 level (1-tailed)
* Correlation is significant at the 0.05 level (1-tailed)

Table 6.18 Means and standard deviations of all organizational variables

No	Organizational variables	Mean (m)	Std. deviation	t Value	Sig. (2-tailed)
OV1 Organizational structure					
1	OV1.1 A flat hierarchy structure	4.6304	0.48802	22.659	0.000
2	OV1.2 A highly bureaucratic top-down structure	3.5217	1.24256	2.848	0.007
3	OV1.3 An ad hoc decentralized structure	3.7609	0.60313	8.556	0.000
4	OV1.4 A routine administrative structure	2.1304	0.74859	−7.878	0.000
OV2 Organizational culture					
5	OV2.1 Learning environment encouraging knowledge sharing	4.5435	0.50361	20.787	0.000
6	OV2.2 Collaborative culture fostering interaction and relationship	4.2174	0.62939	13.119	0.000
7	OV2.3 Open-mindedness climate encourages mutually trusting and honest interaction	4.3696	0.74113	12.533	0.000
8	OV2.4 Understand and commit to a shared vision	4.1304	0.65349	11.732	0.000
9	OV2.5 Crisis-prepared mindset	4.4130	0.6856	13.979	0.000
10	OV2.6 Develop crisis goal alignment	4.6087	0.49344	22.112	0.000
11	OV2.7 Facilitate a climate of constant change, conflict, and diversity	3.1522	0.96534	1.069	0.291
12	OV2.8 Encourage continuous discourse of different points of view	3.1087	0.79522	0.927	0.359
OV3 Information technology capability and information management system					
13	OV3.1 Usage of intranet within the company with e-mail communication system	4.4130	0.65238	14.690	0.000
14	OV3.2 A dedicated crisis or emergency website in the company's intranet system	3.6739	0.92025	4.967	0.000
15	OV3.3 Usage of proper modern information technology to facilitate communication	4.4565	0.62206	15.881	0.000
16	OV3.4 Emergency phone numbers are posted and clearly marked beside all site phones	4.0435	0.63093	11.217	0.000
17	OV3.5 Adopt an innovative integration information system	4.3696	0.67852	13.690	0.000
18	OV3.6 Proper instructing information available	4.4783	0.50505	19.852	0.000
19	OV3.7 Feedback and response to the issues are prompt in a timely fashion	4.4739	0.47396	23.954	0.000
OV4 Management and leadership style					
20	OV4.1 A flexible communication management process and procedures	4.4092	0.69234	13.112	0.000
21	OV4.2 A rigid and centralized communication management process and procedures	3.1739	1.08124	1.091	0.281

(continued)

Table 6.18 (continued)

No	Organizational variables	Mean (m)	Std. deviation	t Value	Sig. (2-tailed)
22	OV4.3 Contractual rules and procedures are strictly followed	3.1087	0.64043	1.151	0.256
23	OV4.4 Project managers or leaders are involved in great details in decision making	2.7826	0.72765	−2.026	0.049
24	OV4.5 A participative decision-making process (democratic)	2.9565	0.91788	−0.321	0.749
25	OV4.6 A decentralized decision-making process	3.6087	0.53658	7.694	0.000
26	OV4.7 A delegation of decision authority style (empowerment)	4.6087	0.49344	22.112	0.000
27	OV4.8 A proactive decision-making process	3.5435	0.62206	5.926	0.000
28	OV4.9 Most decision are made through consensus	2.4565	0.78050	−4.723	0.000
29	OV4.10 Critical decisions are made by some dominant elites	3.5000	0.72265	4.693	0.000
30	OV4.11 An influencing, motivating, and inspiring leadership style	4.5652	0.54374	19.524	0.000
31	OV4.12 A control of authority leadership style (centralized)	3.1087	0.82269	0.896	0.375
32	OV4.13 A risk-taking preference leadership style (over-confident)	1.8913	0.52613	−14.292	0.000
33	OV4.14 A habit-path resort preference leadership style (empiricism)	3.0217	0.77428	0.190	0.850
34	OV4.15 A long-term decision-making vision	3.8913	0.48204	12.541	0.000
35	OV4.16 A constructive conflict management for a well-managed conflict	3.8478	0.36316	15.834	0.000
36	OV4.17 A conflict avoidance and elimination preference	3.0217	0.25726	0.573	0.569
37	OV4.18 Encourage a defined sense of purpose	4.4565	0.62206	15.881	0.000
38	OV4.19 Crisis management plan (if applicable) is strictly followed and given credit to	3.4783	0.50505	6.423	0.000
39	OV4.20 Expertise to deal with crisis contingency	4.2174	0.51264	16.106	0.000
40	OV4.21 Ability to influence over others	4.3696	0.64494	14.403	0.000
OV5 Sense-making capability of organizations and members					
41	OV5.1 High level of trust when communicating openly and truthfully	4.3913	0.68242	13.828	0.000
42	OV5.2 Offer help and provide reciprocal service	3.1087	0.84927	0.868	0.390
43	OV5.3 Show respect to one another, and being accountable to each other	4.4783	0.50505	19.852	0.000
44	OV5.4 Confident and show a positive attitude to the project organization	4.4871	0.49782	21.621	0.000
45	OV5.5 Transparency of intent and straightforward actions	4.5352	0.58525	17.887	0.000
46	OV5.6 Strong feeling of obligation and friendship	2.0000	0.69921	−9.700	0.000

(continued)

Table 6.18 (continued)

No	Organizational variables	Mean (m)	Std. deviation	t Value	Sig. (2-tailed)
47	OV5.7 Problems can be recognized at the very first sign and dealt with quickly	4.0217	0.44667	15.514	0.000
48	OV5.8 Aware of their risks and obligations as well as others	3.8913	0.60473	9.996	0.000
49	OV5.9 Defensive mechanism strategies to avoid anxiety	3.0217	0.57693	0.256	0.799
50	OV5.10 Self-reflection about the crisis situation and willing to engage in double-loop learning	4.4871	0.49782	21.621	0.000
51	OV5.11 Willing to raise issues and take risks	3.8261	0.48554	11.539	0.000
52	OV5.12 Willing to accept constructive feedback	3.5435	0.50361	7.319	0.000
53	OV5.13 Maintain a good interpersonal working relationship	3.8478	0.75916	7.574	0.000
OV6 Skill capability of members of the organization					
54	OV6.1 Good communication skills to discourse and convey their opinions	4.1304	0.58152	13.184	0.000
55	OV6.2 Capability of coping and resolving differences in opinions	3.5652	0.50121	7.649	0.000
56	OV6.3 Good problem-solving and analysis skills to handle unforeseen circumstances.	3.7391	0.49147	10.200	0.000
57	OV6.4 Capability to cope with pressure inherent in crisis situations	4.0435	0.46935	15.079	0.000
58	OV6.5 Functional knowledge to deal with paradox and ambiguities inherent in crisis situations	4.0652	0.38885	18.579	0.000
59	OV6.6 Competent technical skills and expertise that would be difficult to replace	3.4130	0.58027	4.828	0.000

Table 6.19 Most important organizational variables orders

No	Organizational variables	Mean	Std. deviation	Rank
1	OV1.1 A flat hierarchy structure	4.6304	0.48802	1
26	OV4.7 A delegation of decision authority style (empowerment), employees have autonomy to make on-the-spot decisions during the crisis	4.6087	0.49344	2
10	OV2.6 Develop crisis goal alignment	4.6087	0.49344	2
30	OV4.11 An influencing, motivating, and inspiring leadership style	4.5652	0.54374	3
5	OV2.1 Learning environment encouraging constantly to acquire and communicate knowledge and share critical crisis information	4.5435	0.50361	4
45	OV5.5 Project participants demonstrate transparency of intent and straightforward actions	4.5352	0.58525	5

6.6.6.1 A Flat Hierarchy Structure

Organizational theory holds that the organizational structure ought to match the demands of the task and environment (Burton and Obel 2004). As identified in the complexity-informed framework, the organizational structure was highly related to various control parameters which influenced the crisis communication system. A flat hierarchy organizational structure was ranked by the respondents as the first most important organizational variable to influence crisis response communication management. When an organization was threatened by a crisis event, the information flowed much higher, communication and controls were expected to flow more horizontally than vertically (Loosemore 1998). Rigid and high-level hierarchical organizational structure was limited by slowing down the information flow rate and increasing the information processing redundancy. It is suggested that although the formal organizational structure was of importance to crisis communication, professionals should supplement the formal organizational structures by creating ad hoc, informal network of communication, through which information might flow more easily and flexibly. The set-up of a crisis management team can increase the interaction of the members across different professional specialties and function sections, through which information may flow more easily and flexibly.

6.6.6.2 A Delegation of Authority Leadership Style

The authority delegation leadership style was ranked as the second most important variable. An adaptive communication system required a particular style of management and leadership, which could be characterized by empowering the authority and responsibility. Information overload, channel bottlenecks, or even break down was considered as the common challenges for crisis communication management during the crisis response stage. Hence, top management should be capable of delegating certain level of decision-making authority, and allowing new processes to emerge freely within certain guidelines at local levels.

6.6.6.3 Development of Crisis Goal Alignment

Development of crisis goal alignment was also ranked as the second most important variable. This variable was identified as one of the significant organizational variables which affected the level of CP3: levels of diversity among agents of the system. There existed high level of diversity in terms of professional specialization and division of labor for the communication system of a construction organization. As espoused in complexity theory, excessive diversity caused instability and chaos. To function effectively, crisis goal alignment was critical to facilitate coordination and integration between the professionals. Project participants would need to learn from each other's behavior in order to mitigate the risks of conflict and misunderstanding in communication. Through the face-to-face interview with some of the respondents, it was highlighted that the development of a common crisis goal should be beneficial to crisis communication management.

6.6.6.4 Influencing, Motivating, and Inspiring Leadership

The third most important organizational variable was an influencing, motivating, and inspiring leadership style. This organizational variable was identified as one of the significant variables affecting CP4: Contained anxiety and tension. Exposure to unexpected and extremely stressful crisis events often increased psychological pressures on the individuals and provoked anxiety in the individuals. Managers could create conditions within the organization that encouraged project members to explore alternatives, restructure around the current issues, and tolerate ambiguity and paradox. A good leader can help the project members to hold and lower their level of anxiety in the face of uncertainty, for example creating the opportunity to discuss the earthquake events in an emotionally safe way and to provide crisis-coping advice, or devised communication strategies tailored to address personal concerns of employees. The findings were consistent with Complexity Theory which espoused that improvement of leadership quality was considered as one of the means to control the anxiety level (Stacey 1996).

6.6.6.5 Learning Environment

A learning environment of constantly encouraging an employee to acquire and communicate knowledge and to share critical crisis information was ranked as the fourth most important organizational variable to influence crisis response communication management. Communication based on the sharing of strategic information was critical to help facilitate project members to make better decisions and become more responsive to the uncertainty and change (Zhong and Low 2011). With this learning environment within an organization, project participants were better able to develop problem-solving mechanisms and to provide reciprocal services for each other, thus helping with coordination and integration for the crisis event.

6.6.6.6 Transparency of Intent and Straightforward Actions

The fifth most important organizational variable was transparency of intent and straightforward actions. In the event of crisis, an increased information load and decreased communication channel often required the project members to share and exchange updates of the crisis situation in a quick and accurate fashion. As mentioned by some of the respondents, the project members' abilities to speak clearly and transparency of intent could prevent misunderstanding and greatly improved the success of crisis management.

6.6.7 Variation Analysis

This section presents a combined analysis of the responses collected from the LSE firms and SME firms. In this research, the categorical variable was defined based on the construction firm's financial qualification class. The purpose of conducting rank correlation analysis using this categorical variable was to evaluate whether the ranking of the organizational variables from different firms (SME vs. LSE) was similar.

The focus of this research was to evaluate the significance of organizational variables to a construction firm's internal organizational communication in response to the earthquake crisis. Therefore, it was inevitable to consider the views of firms under different categories on the significance of the organizational variables. Take for instance, a LSE can have a more complicated communication structure while a SME has a relatively simple structure. The research investigated the responses of these two companies in the event of an earthquake.

Separate calculations of the mean importance and standard deviations were made for responses received from the SME firms and LSE firms. The organizational variables were ranked separately in their order of importance as shown in Table 6.20

From Table 6.20, the SME firms and LSE firms generally gave similar rankings to most of the organizational variables. However, they varied their views on some of the organizational variables, which are discussed below:

- For organizational structure related variables, SME respondents recognized the importance of OV2.1, a flat hierarchy structure, to effective crisis

Table 6.20 Organizational variables ranked by mean importance ratings

Organizational variables	SME			LSE		
	Mean	Std. deviation	Rank	Mean	Std. deviation	Rank
OV1 Organizational structure						
OV1.1 A flat hierarchy structure	4.3684	0.49559	18	4.8148	0.39585	1
OV1.2 A highly bureaucratic top-down structure	4.7368	0.45241	1	2.6667	0.83205	53
OV1.3 An ad hoc decentralized structure	3.8421	0.60214	32	3.7037	0.60858	34
OV1.4 A routine administrative structure	2.1579	0.68825	57	2.1111	0.80064	57
OV2 Organizational culture						
OV2.1 Learning environment encouraging constantly to acquire and communicate knowledge and share critical crisis information	4.5789	0.50726	7	4.5185	0.50918	13
OV2.2 Collaborative culture fostering interaction and relationship between project participants	4.2105	0.53530	20	4.2222	0.69798	22
OV2.3 Open-mindedness climate encourages mutually trusting and honest interaction	4.5263	0.51299	10	4.2593	0.85901	20
OV2.4 Understand and commit to the organization's vision	4.1053	0.65784	22	4.1481	0.66238	26
OV2.5 Employees are fully aware of the emergency communication procedures and have crisis-prepared mindset	4.1579	0.76472	21	4.5926	0.57239	6
OV2.6 Develop crisis goal alignment	4.6842	0.47757	3	4.5556	0.50637	8
OV2.7 Facilitate a climate of constant change, conflict, and diversity	3.3684	0.95513	44	3.7037	0.46532	34
OV2.8 Encourage continuous discourse of different points of view	3.3158	0.58239	45	3.6630	0.89792	37
OV3 Information technology capability and information management system						
OV3.1 Usage of intranet within the company with e-mail communication system	4.1053	0.73747	22	4.6296	0.49210	4
OV3.2 A dedicated crisis or emergency website in the company's intranet system	2.8421	0.76472	53	4.2593	0.44658	20
OV3.3 Usage of proper modern information technology to facilitate communication	4.5263	0.77233	10	4.4074	0.50071	16
OV3.4 Emergency telephone numbers and the site location are posted and clearly marked beside all site telephones	3.8947	0.45883	29	4.1481	0.71810	26
OV3.5 An innovative integration information system is adopted	3.7368	0.45241	38	4.8148	0.39585	1
OV3.6 Employees have proper instructing information they need to know to protect themselves physically	4.6316	0.49559	4	4.3704	0.49210	17
OV3.7 Feedback and response to the issues are prompt and in a timely fashion	4.4316	0.49559	13	4.5037	0.46532	12

(continued)

Table 6.20 (continued)

Organizational variables	SME			LSE		
	Mean	Std. deviation	Rank	Mean	Std. deviation	Rank
OV4 Management and leadership style						
OV4.1 A flexible communication management process and procedures to allow for open communication	4.5891	0.41890	2	4.2811	0.50920	19
OV4.2 A rigid and centralized communication management process and procedures	4.0000	0.66667	25	2.5926	0.93064	54
OV4.3 Contractual rules and procedures are strictly followed during the response stage	3.2632	0.73349	46	3.0000	0.55470	51
OV4.4 Project managers or leaders are involved in great details in decision making	3.0526	0.70504	49	2.5926	0.69389	54
OV4.5 A participative decision-making process (democratic)	2.5263	0.51299	56	3.2593	1.02254	45
OV4.6 A decentralized decision-making process	3.6842	0.58239	39	3.5556	0.50637	41
OV4.7 A delegation of decision authority style (empowerment), employees have autonomy to make on-the-spot decisions during the crisis	4.5789	0.50726	7	4.6296	0.49210	4
OV4.8 A proactive decision-making process	3.5263	0.61178	41	3.5556	0.64051	41
OV4.9 Most decision are made through consensus	2.6316	0.59726	55	2.3333	0.87706	56
OV4.10 The critical decisions are made by some dominant elites who have the power to delegate decisions	4.0000	0.57735	25	3.1481	0.60152	47
OV4.11 An influencing, motivating, and inspiring leadership style	4.6316	0.59726	4	4.5185	0.50918	9
OV4.12 A control of authority leadership style (centralized)	3.2105	0.71328	47	3.0370	0.89792	49
OV4.13 A risk-taking preference leadership style (over-confident)	2.1053	0.56713	58	1.7407	0.44658	59
OV4.14 A habit-path resort preference leadership style (empiricism)	3.0000	0.81650	50	3.0370	0.75862	49
OV4.15 A long-term decision-making vision	3.7895	0.53530	35	3.9630	0.43690	30
OV4.16 Constructive conflict management for a well-managed conflict	3.8421	0.37463	32	3.8519	0.36201	32
OV4.17 A conflict avoidance and elimination preference	2.9474	0.22942	52	3.0741	0.26688	48
OV4.18 Encourage a defined sense of purpose	4.4737	0.61178	12	4.4444	0.64051	14
OV4.19 Crisis management/communication plan (if applicable) is strictly followed and given credit to	3.6316	0.49559	40	3.3704	0.49210	44
OV4.20 Manager's expertise to deal with crisis contingency	4.4211	0.60698	14	4.0741	0.38490	28

(continued)

Table 6.20 (continued)

Organizational variables	SME			LSE		
	Mean	Std. deviation	Rank	Mean	Std. deviation	Rank
OV4.21 Manager's ability to Influence over others	4.4211	0.76853	14	4.3333	0.55470	18
OV5 Sense-making capability of organizations and members						
OV5.1 Project participants feel a high level of trust when communicating openly and truthfully	4.1053	0.80930	22	4.5926	0.50071	6
OV5.2 Project participants offer help and provide reciprocal service	2.9474	0.77986	52	3.2222	0.89156	46
OV5.3 Project participants show respect to one another, and are accountable to each other	4.4211	0.50726	14	4.5185	0.50918	9
OV5.4 Project participants are confident and show a positive attitude to the project organization	4.5316	0.49559	9	4.4556	0.50637	13
OV5.5 Project participants demonstrate transparency of intent and straightforward actions	4.3684	0.59726	18	4.6667	0.55470	3
OV5.6 Project participants have a very strong feeling of obligation and friendship	1.9474	0.70504	59	2.0370	0.70610	58
OV5.7 Problems can be recognized at the very first sign and dealt with quickly, without transfer of blame to those who are not responsible	3.9474	0.52427	28	4.0741	0.38490	28
OV5.8 Our employees are fully aware of their risks and obligations as well as others	3.8947	0.65784	39	3.8889	0.57735	31
OV5.9 Develop defensive mechanism strategies to avoid anxiety	3.0000	0.66667	50	3.0370	0.51750	48
OV5.10 Our employees tend to self-reflect about the crisis situation and are willing to engage in double-loop learning	4.5842	0.47757	6	4.4185	0.50918	15
OV5.11 Our employees are willing to raise issues and take risks	3.8421	0.60214	32	3.8148	0.39585	33
OV5.12 Our employees are willing to accept constructive feedback	3.4737	0.51299	43	3.5926	0.50071	38
OV5.13 Our employees maintain a good interpersonal working relationship	4.4211	0.60698	14	3.4444	0.57735	43
OV6 Skill capability of members of the organization						
OV6.1 Our employees possess good communication skills to discourse and convey their opinions	4.0000	0.66667	25	4.2222	0.50637	22
OV6.2 Our employees are capable of coping and resolving differences in opinions	3.5263	0.51299	41	3.5926	0.50071	38
OV6.3 Our employees possess good problem-solving and analytical skills to handle unforeseen circumstances	3.7895	0.41885	35	3.7037	0.54171	34
OV6.4 Our employees possess the capability to cope with pressure inherent in crisis situations	3.7895	0.41885	35	4.2222	0.42366	22

(continued)

Table 6.20 (continued)

Organizational variables	SME			LSE		
	Mean	Std. deviation	Rank	Mean	Std. deviation	Rank
OV6.5 Our employees possess functional knowledge to deal with paradox and ambiguities inherent in crisis situations	3.8947	0.31530	29	4.1852	0.39585	25
OV6.6 Some of our employees possess competent technical skills and expertise that would be difficult to replace	3.1579	0.50146	48	3.5926	0.57239	38

communication management. However, they ranked OV1.1, a highly top-down structure, as the most important variable, while LSE viewed this organizational variable as among the last few least important variables. This large disparity in ranking might indicate that SME firms with relatively small-scale projects preferred a centralized structure when responding to the crisis.

- Similarly, SME firms also ranked OV4.2, centralized communication management process and procedures, as a more important variable, while LSE firms regarded this as less significant.
- In terms of organizational variables under the information technology capability and management system category, SME firms ranked OV3.2 and OV3.5 as less significant variables, while LSE firms put more weight on both variables. This appears to imply that LSE firms placed more emphasis on the adoption and deployment of innovative information technology such as dedicated crisis management website and integration information system to facilitate crisis communication.
- While SME firms ranked OV5.13, good interpersonal relationship, as the14th most important variables, the LSE firms regarded it as of no significance. This might be because SME firms relied more on the employee's good interpersonal relationships to facilitate interaction and communication.

Variation analysis was conducted to examine whether the ranking of the 59 organizational variables from the SME firms and LSE firms were correlated. The Spearman's rank correlation technique was applied. As shown in Table 6.21, the Spearman's rank correlation results indicated that the rankings by the SME firms and LSE firms were strongly correlated. It was therefore confirmed that the respondents generally agreed on the significance of the organizational variables identified for crisis response communication management in the building projects in Chengdu.

Table 6.21 Spearman's rank correlation for organizational variables

			SME firms	LSE firms
Spearman's rho	SME firms	Correlation coefficient	1.000	0.693**
		Sig. (2-tailed)	.	0.000
		N	59	59
	LSE firms	Correlation coefficient	0.693**	1.000
		Sig. (2-tailed)	0.000	
		N	59	59

** Correlation is significant at the 0.01 level (2-tailed)

6.7 Case Studies

It was necessary to further analyze and explain the findings from the quantitative studies in the previous sections. In the next, the conceptual framework was applied and validated with two case studies on two construction firms, respectively. The case studies were able to present a more holistic and in-depth understanding of the dynamics of crisis response communication practice in the Chengdu construction firms, the emergent communication pattern, the decision-making processes, and communication strategies during the critical period of the 2008 Sichuan earthquake. In particular, it would be necessary to consider the impact of the significant organizational variables on the organization's ability to adapt flexibly to the contingencies and discontinuities, and whether the organizations were able to influence the control parameters by adjusting their organizational variables, and thus to make connections with the principles of the complexity theory.

6.7.1 Organizational Communication Structure Analysis

The analysis is presented in the following Sects. 6.7.1.1–6.7.1.4.

6.7.1.1 Case Study Methodology

As discussed in Sect. 6.6.6, one of the most important organizational variables that were taken into accounts by Chengdu's construction firms in crisis response communication management was the organizational communication structure which was expected to be able to hold and exchange valid information through all parts of the system. The case studies further discussed organizational communication structure of this Chengdu's construction firm in response to the earthquake, and analyzed the complex interactions and the process of self-organization among the organizational members at a local level.

A State-owned LSE construction firm was selected for the case study. The firm has qualification of Class C1 for contracting highway engineering construction projects, municipal public works and general building engineering projects. At the time of the 2008 Sichuan earthquake, the company had in total 24 building projects underway throughout the Sichuan province.

The information was collected through a series of structured interviews with the project manager and the manager of the production department involved with the response to the earthquake. The interviews were tape-recorded with the permission of the interviewees and fully transcribed subsequently.

6.7.1.2 Emerging Communication Patterns

At the outset of the earthquake on May 12 2008, a Crisis Management Team (CMT) or Crisis Task Force (CTT) was swiftly set-up in the company's headquarter for implementing the Crisis Management Plan (CMP) and coordinating the crisis response. The CMT was headed by the company general manager and management representatives from the department of production and the department of safety. The organizational communication structure followed a two-level hierarchy as illustrated in Fig. 6.2.

At the company level, the CMT promptly communicated with the project managers from different project worksites to collect important project information and to disseminate critical command and crisis decisions from the headquarter. The project managers played significant roles in terms of feeding and reporting information that was important for the CMT.

At the project level, an emergency response team which was led by the project manager was also assembled to execute the CMP procedures. The emergency response team at each of the construction worksite was further divided into different sub-teams with different emergency-related functions. Take for instance, the site rescue team led by the field supervisor was composed of the foremen of different sections, e.g., reinforcement, concrete, mould, and machinery. The site rescue team was responsible for identifying and reporting potential hazards and risks, rescuing injured workers, and protecting on-site materials and equipment from serious damages. The medical treatment team, equipped with first aid trained staff and workers, was in charge of providing assistance to on-site medical emergency cases. The defense safety team with designated staff from the safety, quality, and security departments took responsibility for evacuation procedures, while the logistics team was in charge of emergency resource deployment, on-site facility maintenance, and fleet-equipment/transportation.

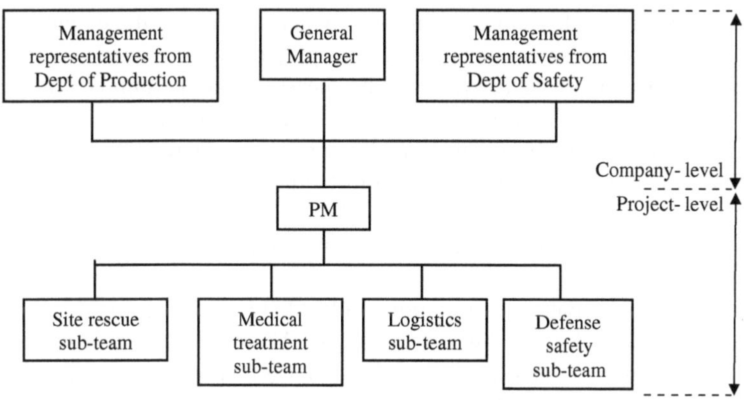

Fig. 6.2 Emergent crisis response communication structure (Case study 1)

The project manager had the responsibilities to communicate and coordinate the works of different emergency response teams/sub-teams, obtained sufficient information regarding the ongoing crisis events, and provided decisions and strategies to minimize the damage to the on-going projects and to the surrounding communities as well.

6.7.1.3 Centrality Characteristics

In terms of the communication structure at the company level, a relatively high centralized characteristic of the organizational communication structure reflected the extent to which information flow in the communication network was centered on the CMT. The CMT was in charge of the overall coordination and crisis decision-making for all the 24 construction projects. This structure seemed to reduce the uncertainties and ambiguities and provided a greater sense of common direction for the crisis response. The construction firm would benefit from having a restricted source but yet widespread supply of information.

Mackenzie (1966, p. 17) defined sociocentric centrality as "the degree to which information flows are centered on one or a few organizational units." The concept of centrality was important because of its potential influence on problem-solving efficiency and behavior. Leavitt (1951) and Shaw (1954) argued that the influence of centrality was dependent on the nature of the task. More complex problems demanded less centralized structures. If an inappropriate degree of centralized control was applied in a situation which demanded flexibility, then the behavioral response would be a defensive one.

This centrality characteristic proved to be important particularly during the initial phase of the earthquake, where the pressure of the situation forced people to throw away the bonds of normal multi-level hierarchy and communicate directly to a source center. As a result of not having to follow routine procedures of communication, information was transmitted rapidly between those involved.

In addition, the communication structure at the company level also presented characteristics of a highly closed centrality, which referred to the extent to which one agent CMT is close to all other agents (project site managers) in the crisis communication network. Closeness can be measured by the number of communication links between two agents. The advantage of this structure was that it produced a tightly-knit communication structure (Loosemore 1998) where people were able to communicate directly without having to go through intermediaries. More directness in communication appeared to speed-up the response process and prevented distortions from occurring in communication as a result of biased intermediaries.

However, a high centrality structure would likely cause information overload, information bottlenecks, and high stress levels. There was a danger of the CMT and/or the command center becoming flooded with information in a very short period of time to resolve different aspects of crisis responses from project managers in different project sites. The problem could be exacerbated in that the crisis managers could experience great difficulties in managing the large flows of information, and fell into an increasingly reactive style of management as information demand ran ahead of its supply. Thus, it would seem that a balance of control and flexibility was needed during a crisis.

It was also found from the emergent communication structure in Fig. 6.2 that the project managers as the focus of information supply within the centrality-driven communication structure occupied critical positions in maintaining free and open information flow. They had positive impact to communication efficiency if they held very positive attitude toward their projects, and expressed the capacities of manipulating the information flow to serve the interests of the projects. In essence, under these conditions, the organization can be strengthened by the positive influence of the occupants of these focal positions.

6.7.1.4 Interdependence and Connectivity

At the project construction site, as shown in Fig. 6.2, all communications from different emergency response sub-teams should have been addressed to the project manager since the project manager was responsible for making decisions about the ongoing crisis activities and evaluating potential options for action. As the central point or hot spot of communication, the project manager should have then taken the responsibility to forward each communication, with his/her additional comments, to the recipient, as appropriate.

To avoid having the communication flow being bogged down, the project team encouraged decentralized and direct flow of communication among the ERT members. The flat and flexible communication flow increased the number of interactions among the team members and allowed them to identify and resolve issues in real time. In order to achieve flexibility and creativity, more fluid communication structure which involved horizontal communication across different professional specialties and functional sections emerged and replaced the traditional hierarchical structures. Strict differentiations between functions in an organization could cause fragmentation. Subsystems should rather be more flexible with an interdisciplinary approach of working together to achieve common organizational goals in crisis management.

Hence, project managers could fulfill the bridging functions and facilitate interaction and network building, as well as contributing to management. From the complexity perspective, management attentions should be placed on the relationships between entities, as well as on the characteristics of the entities themselves.

Traditionally, the interpretation of data and information was done by top management, which in turn led to subjectivity, exclusivity, and over-control. Complexity theory espoused that attention should be given to the interdependencies and connectivity between the agents, groups, and/or different sub-systems in an organization. The interdependence suggested that all the sub-systems and agents should take part in the process of the systems. The participation could add to the richness of information, shared responsibility, more trust, and ultimately to healthier relationships. This interdependency and participation in turn inferred relationships: the sharing in decision making, as well as in the dissemination and interpretation of information throughout the organization.

In addition, the communication network at the project level which collected information widely, through a multitude of alternative paths and channels, provided flexibility in communication and avoided distortion. As a matter of facts, with multiple agents or units reviewing and checking the progress of the crisis events and response processes as these unfolded, errors and areas for reinforcement would be discovered more readily. While the number of elements helped the system to register different kinds of inputs, the independence of the elements prevented these from being affected too much by the activities of others. Both features stimulated sensitivity to and perceptiveness of the object of attention.

6.7.2 Tai Fa Construction Engineering Co., Ltd

In this section, the complexity-informed crisis response communication model are applied and tested on a local construction firm—Sichuan Tai Fa Construction Engineering Co., Ltd. This section analyzes the crisis response communication system of the construction firm from a complex system's perspective, and assesses the adaptive capability and flexibility of the company in response to the Sichuan earthquake.

In the first instance, the construction firm, its organizational communication structure, and the earthquake response practice are briefly reviewed, followed by the introduction of the case study methodology. Then, the communication challenges and the decision-making processes experienced during the response stage of crisis management are examined. Next, how the company and the crisis management team successfully created a collaborative environment that supported its adaptation during the response stage and the communication strategies adopted are analyzed. And lastly, some final remarks about the adaptability and innovations that had been created throughout the earthquake response stage are concluded in this section.

6.7.2.1 Introduction to the Company

Sichuan Tai Fa Construction Engineering Co., Ltd was founded in 1993 as a privately-owned enterprise. During the period of this fieldwork in 2009, this company had total assets of 23 million RMB Yuan and possessed the Grade 2 general contracting qualification, sanctioned by the Ministry of Construction of China, for building construction and municipal engineering works. The company offered a complete construction service for a wide range of project types in the Sichuan province, including commercial and residential properties. Its total contracting revenue in 2009 was 500 million RMB Yuan. By the end of 2010, the total assets of this company were increased to 50 million RMB Yuan and the company successfully acquired the Grade 1 general contracting qualification for building construction.

The company followed a flat hierarchical organizational structure at the company level and consisted of six departments directly administered by the top executive management in its headquarter, as shown in Fig. 6.3. At the project level, the company adopted the project manager liability system, where the project managers as the authorized corporate legal representatives were fully responsible for construction management and operational issues at the construction site and communication with headquarter in a routine fashion.

The company had incorporated the hazard analysis and safety construction principles in every phase of the construction and had a GB/T19001-2000/ISO9001:2000 certification awarded by the Beijing China Construction Certification Center. During the period of this fieldwork in early 2009, the company was audited and awaiting certification to the GB/T19001-2008/ISO9001:2008 standard. During the last few years, the company had steadily enhanced its quality management and safety climate, which was to some extent occasioned by the

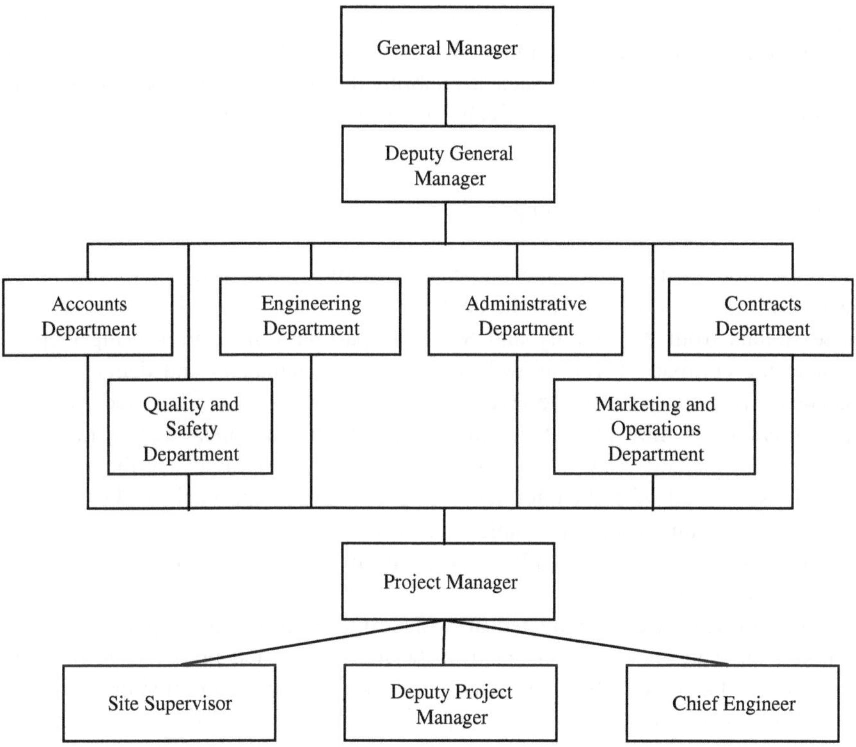

Fig. 6.3 Organizational structure (Case study 2)

Chinese Government's enforcement of safety regulations, as well as the improvement in nationwide safety awareness level of the public. For example, the "Administrative Regulations on the Work Safety of Construction projects" (Decree No. 393 of the State Council of the People's Republic of China) came into force on February 2004. It introduced a marked change in the health and safety provisions in relation to construction. Any noncompliance with the regulation or accident hazards would lead to an investigation and ultimately to an appropriate penalty accordingly.

In 2005, the company issued a series of documents on the management methods of production safety. Construction safety inspections at each construction site were initiated by the top management, and conducted by the safety officers and safety committee representatives from the head offices. The company had adopted health and safety auditing as its major focus to assist employees in meeting their duty of care obligations. To augment safety and quality education as well as professional skills of different trade employees, the company had promoted the worker's night school opportunities at every project construction site since 2005.

Over several years of its operations, the company had developed Crisis Management Plans covering a wide of range of specific site scenarios: suspending/securing crane operations, utility collapse, electrical shock, fire emergency, environmental pollution, and food poisoning. In terms of natural disasters, e.g., flood, storm, hurricane, earthquake, the company had provided general guidelines for the emergency operations, rather than specific response planning. Those programmes and procedures were seen as enabling the company for both management and the employees to respond actively in various emergency situations.

6.7.2.2 Case Study Methodology

The case study data were collected mainly in two steps. First, a formal set of interviews were conducted with the deputy general manager, senior managers, and professionals from the Quality and Safety Department and Engineering Department at the company level, as well as the project managers and deputy project managers from the project site levels. The interviewees were identified based on the critical roles they played in communication management during the 2008 Sichuan earthquake; they were themselves directly involved in the response operations. A total of eight interviews were completed and each of the interviewees' roles is summarized in Table 6.22. Due to the sensitivity of the personal information provided, anonymity has been maintained in this study without revealing their names.

The interviews were conducted at a place the participants suggested, usually their offices. The interviews lasted from 60 to 90 min, and involved an in-depth one-to-one discussion between the managers/senior professionals and the researcher. A few rounds of follow-up telephones and e-mail discussions with some participants were also made to further confirm particular points in the response communication process. The majority of the respondents (Six out of

Table 6.22 Summary of interviewee's role

No.	Role
1	Deputy manager
2	Department manager from Quality and Safety Department
3	Senior engineer from Quality and Safety Department
4	Senior manager from Engineering Department, in charge of logistics and materials
5	Chief engineer from Engineering Department, in charge of construction technology
6	Project manager from Project A
7	Project manager from Project B
8	Deputy project manager from Project C

eight) had worked for the company for more than 10 years; thus they were experienced, could contribute valuable information and identified the significant issues that influenced communication performance during the crisis.

The respondents were asked a consistent series of questions regarding the aforementioned organizational variables, as well as their decision-making and communication strategies related to the critical period in response to the earthquake. The researcher also invited the respondents to describe the events associated with the crisis and to evaluate the organizational efforts and overall organization's response to it from their own perspectives. This provided the respondents' first-person accounts and narratives of their experiences in relation to the crisis response practices.

All of the interviews were audio-taped and transcribed by the researcher afterward. Apart from recording, the researcher also took notes for each interview. The transcripts of the narratives and notes served as the main document resource of the analysis.

Second, the Crisis Management Plans, internal safety production documents, and a series of documents and statements during the intensive period of the earthquake were reviewed and examined, which were made available from the generous support and cooperation of the company. All of these sources were used to provide a rich perspective, holistic, and dynamic view of the organization's crisis response communication practice, and to assess the adaptive capability of the organization in response to the 2008 Sichuan earthquake crisis.

6.7.2.3 Communication Challenges Faced

The telecommunication infrastructure throughout the Sichuan Province was severely affected and half of the wireless communications were lost in Sichuan during the first few hours following the earthquake. One of the most salient qualities of this crisis was a lack of timely and reliable information about the ongoing crisis activities at the company's construction project jobsites, as to what had happened, and if anything, what the project site team could do to resolve the problem. Due to the inability to use the normal channels of communication, e.g., telephones, mobile phones, short messaging services, the project site teams also faced difficulties in reporting the site activities back to the headquarter in the first place.

Message overload was another serious problem for the crisis management team attempting to make sense of a pile of data and to extract relevant information from different project sites. The crisis team members should be able to make decisions rapidly because the crisis situation did not afford them the luxury of spending hours around a conference table discussing the pros and cons of each step and then to take a vote. The stress created by the accelerated pace, coupled with an information vacuum at the outset of the earthquake, generated increased anxiety for the team members.

In the project jobsites, the workforce exposed to the massive earthquake which they have not experienced before, exhibited the reactions of fear and stress. The uncertain conditions of the earthquake in combination with the high informational needs might enhance the relative levels of ambiguity. In these cases, the sense-making capability of individuals seeking to manage and contain the crisis may be overwhelming. The accelerated stress might impair their cognitive, emotional, and decision-making abilities. In addition, as some of the workers were from the earthquake epicenter region, they were concerned and desperate for information related to the situation of their family members. In this case, they could be severely distracted by such concerns from attending to the incoming crisis information and concurrently performing jobsite activities. As a result, management of the emergent behaviors during the stressful times was one of the challenges that the crisis teams and project managers faced.

In the sections that follow, we shall see how the organization and crisis management team were able to modify their organizational variables and reorganize their environment in order to facilitate adaptation and self-organization, when faced with those post-earthquake related uncertainties and challenges.

6.7.2.4 Flexible Communication and Decision-Making

At the outset of the earthquake on May 12 2008, a CMT was swiftly set-up in the company's headquarter for implementing the CMP and coordinating the crisis response. The CMT was headed by the company general manager and deputy manager, and its core members included management representatives from the Quality and Safety department and Engineering department, as well as project managers from different construction worksites.

The CMT was one major entity in which company key managers met to communicate and share their expertise to solve the critical crisis issues. The emerging organizational communication structure during the crisis is illustrated in Fig. 6.4.

At the project level, a project emergency response team was also assembled to coordinate and communicate the evacuation procedures and rescue operations in the worksite. The emergency response team composed of the project manager, the

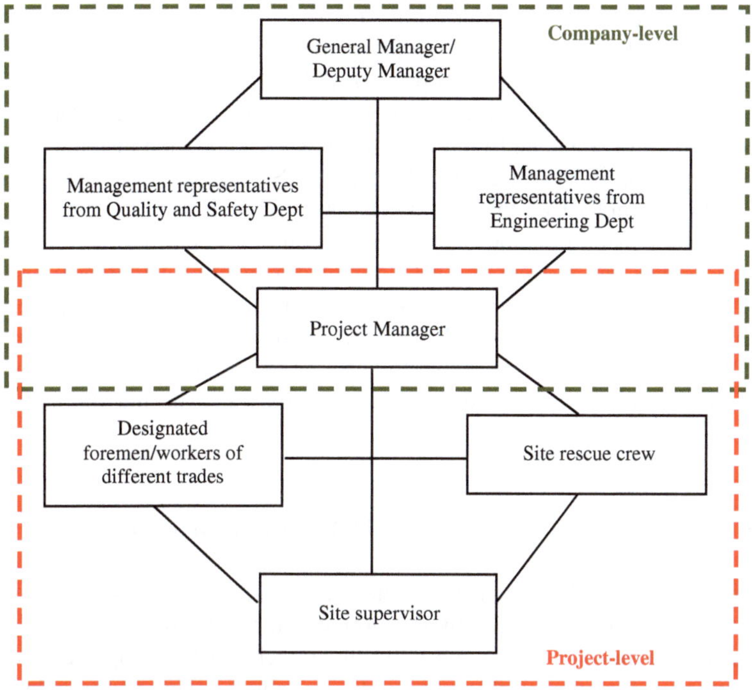

Fig. 6.4 Emergent crisis response communication structure (Case study 2)

site supervisor, the designated foremen or workers from different trades, as well as the rescue crew who was trained in first aid and medical emergencies.

The emerging communication dual-structure at the company level and the project level in order to enhance the flexibility of the process added tremendous value to crisis response management. On the one hand, the flat and decentralized communication flow within each level increased the number of interactions among the team members and allowed them to identify and resolve issues in real time. For instance, direct and horizontal communication across different professional specialties and functional sections in the CMT replaced the traditional routine hierarchical structure. The team members were able to quickly verify their understanding about an issue, instead of making assumptions and rectifying their understanding over lengthy feedback loops. To avoid bogging down the communication flow, the site project team also encouraged decentralized and direct flow of communication among the emergency response team members.

On the other hand, the duality characteristic of the emerging structure facilitated the flow of information between the company crisis management team and the site emergency response team. There was common agreement in the literature that a crisis response system required communication of initial events as early as possible. In the event of critical issues occurring during the earthquake, the site project manager had a ready, responsible respondent with whom to communicate, benefited by the dual communication structure. Furthermore, the CMT was capable of providing expertise and sufficient resources to be utilized in handling these issues and incidents at the operational level. Consequently, it helped to reduce the redundancy between the time when an issue on site was identified and the time needed for someone from the company level to provide insights to resolve the issue, by creating short feedback loops.

It should be emphasized that the effectiveness of such flexible communication structure and decision-making system was not due merely to its pure implementation. Moreover, the cooperative behavior of organizational members and their interactive and collaborative relationship exerted throughout the crisis response stage enabled the organization and crisis management team to flexibly overcome the challenges and to adapt to the earthquake crisis. Next, we will highlight and discuss those characteristics and behaviors demonstrated during the earthquake crisis.

6.7.2.5 Balancing the Degree of Power Differentials Among the Teams

Structure reflects to what extent the management defines the project process and imposes formal control and communication structures. Favorable project performance is not obtained by tight control, nor by anarchy, but by empowerment. The construction organization in this case study was able to successfully balance the degree of power differentials among its organizational members, thus facilitating the coordination of activities throughout the crisis response stage.

It appeared that the site team was able to balance the power differentials among the emergency response team members, by the clear division of the responsibilities, which leveraged on the individual strength and facilitated decision-making processes with regards to the emergency operation on site. This facilitated flexibility for the emergency response team members to act promptly according to their better judgment and deal with the emergencies as they deemed fit appropriately.

Key individuals in the project site have been empowered so that they can make decisions at a critical point in time. For instance, the foremen from different trades had certain level of decision-making power that they did not need to constantly report and check with the project manager; they could expedite the process and minimize the damage by making a decision based on their own Individual empowerment. This was critical to support real-time communication and decision making. Without it, it would have been very difficult for the project team to minimize communication redundancy and delays.

At the company level, faced with limited resources, time pressure, and stress, the CMT members must rapidly understand the status quo crisis situations from all the project jobsites and come to critical decisions quickly. The project managers as the main source for crisis information had enough decision-making power and authority to filter and distribute the project worksite information. Effective communication filter prioritized information relevant to crisis choice, determining which ones were important for the CMT, and which ones could be safely ignored. As explained by the deputy manager,

> During the critical period, the availability of timely and relevant crisis information could facilitate us to handle the deluge of incoming information occurring during the crisis response. We encouraged the site project managers to make decisions based on their experiences and judgment regarding the project site issues which might not be necessary to the upper management level. They didn't need to report and check in repeatedly over every detail. The majority of project managers did very good job by allowing the CMT to focus only on those points salient to the critical issues at hand.

However, empowerment also required key individuals to have enough project management experience and skills to take on such a responsibility. Depending on the project size, decision-making responsibility might need to be delegated to lower level individuals who might not have the expertise required to make the right decisions. In such a case, empowerment might have a negative impact to the project performance.

6.7.2.6 Unleashing the Distributed Intelligence in CMT

The emerging crisis management team was structured for the key management members to meet to share their expertise. To facilitate and unify the amount of coordination, within the first 2 h after the earthquake hit, the company swiftly decided to set-up a crisis response command center in one of its construction worksites nearby the company head office, where there were semi-assembly rooms at lower floors with relatively good aftershock-resistance.

Such a command center served a focal point of responsibility for decision making and information management relating to the earthquake response. It also co-located all the key personnel involved in crisis response management. This transformed the crisis management team into a communication network in which information flowed freely among the diverse experts from different departments in the room.

In the aftermath of the earthquake, organizations needed to gather intelligence efficiently and from the right sources; they also needed to process that information quickly to make sense of the current situation and to come to a decision. The face-to-face interactions and discussions between the team members provided a platform to share their respective points of view to resolve the critical issues at hand. It was also meant to help to minimize the amount of heterogeneity and specialization divergences within the team, thus facilitating integration and collaboration.

As pointed out by the deputy manager,

> Individuals may possess unique expertise and knowledge, but no single person can claim full knowledge on what happened throughout the entire process. The crisis management team played a key role in sharing the expertise and providing the distributed collective intelligence that resided throughout the organizations.

Although no single individual in the crisis team possessed all of the necessary knowledge or skills to manage the crisis, team members drawn from different departments within the company could work together closely over time to provide organizational knowledge and to make decisions on how problems might be resolved, thus saving precious time to the team.

This collective insight of organizational knowledge was supported by the theories of complexity, which emphasized not only the importance of multiple sources of information, but also multiple points of view. In a complex system consisting of many intelligent agents, knowledge was both local and distributed. No agents have complete knowledge about the behavior of the system as a whole; although agents were presumed intelligent, "that intelligence is local to their position on the landscape" (Levinthal and Warglien 1999, p. 345). This local situated knowledge spread incrementally through local interactions between the number of individuals in a complex system. Hence the local, but distributed characteristics of knowledge actually resembled the effects of local complex interactions between the individual agents.

6.7.2.7 Fostering Collaborative Dialog

The collaborative communication environment facilitated crisis response management in both the company level and the project level. During routine operations, the company conducted internal quality audits regularly at the headquarter. The management representatives from the Quality and Safety department had the responsibility to maintain the quality management system and to ensure that quality processes were carried out properly for all the construction worksites. Based on the interviews with the project managers, it was noted that the project managers attended the quality meetings once in a week in the head office to report on the progress of the construction sites, and they were encouraged to discuss issues with management representative from the Quality and Safety department in both formal and informal meetings.

Benefiting from the routine and regular patterns of interactions between the project team and the department of Quality and Safety, they developed a shared intuition that allowed them to make better decision more quickly than with less collaborative management teams. The site project manager became knowledgeable about the requirements from the top management and fully understood the rationale of the decisions made. Regular and collaborative communications between the site and organizational operations was a norm. Communication could be considered as a collective process in which people pooled their expertise, values, and information.

Consequently, communication in response to the extraordinary nature of the earthquake event was facilitated and the emergency issues received early communication that was desired. When resolving the critical emergent issues, the project manager could facilitate the sharing of crisis events and enable the crisis management team to make informed decisions quickly at the early stages of crisis response, which would not need to be re-evaluated at a later stage, thus saving time.

According to the project manager, at the construction site, quality meetings were held on site regularly not only as a forum to communicate general information, but also as a place to raise issues and propose solutions. These meetings acted as team-building sessions for the personnel on site to become more interactive. Negative attitudes most often originated from ignorance or from old habits and mental norms, rather than from intentions. Hence, such attitudes could be changed by individuals who were willing to influence their peers. Through healthy and constructive dialog throughout the project, the team was capable of resolving differences in opinions. Due to the high commitment of each participant to the project vision and goal, the company created a positive environment for resolving differences in opinions.

Eventually as the participants became familiar with the processes and methods of working, trust and a sense of teamwork between the project members emerged and improved. Professional behaviors based on trust and mutual respect could facilitate more collaborative communication. Cooperation among the crisis teams and individuals was critical when operating in complex environments.

Intuitively, the cooperation seemed to have been better in the field than between administrative offices, mostly likely because resolving field issues could not have simply been achieved with endless negotiations and cumbersome paperwork. Field issues that were more straightforward and required immediate actions were resolved via teamwork. As a matter of fact, teamwork seemed to have occurred at the lower levels of each of the emergency response teams, as individuals learned how to deal directly with each other, thus developing trust on a person-to-person basis.

6.7.2.8 Building up Collective Responsibilities

Project B was one of the residential building projects operated by the company in the Chengdu urban area. This project witnessed teams and individuals taking collective responsibilities for coping with the emergency at the construction site. It was viewed as a success in terms of positive team relationships.

Around 2:30 pm on May 12 2008, the earthquake hit when a mechanical worker was operating the tower crane halfway to lift a cooling tower to the roof of the building. Due to the strong tremors consecutively over the next few minutes, the cooling tower broke loose from the rigging, and the tower crane was inclined forward to hit the top floor of the main building with a very loud crash, along with the cooling tower.

This situation was far removed from the comfortable zone within which the project manager and the worksite engineers and workers had been used to operate in. While quickly evacuating the site workers and rescuing the operative out of the inclined tower crane, the project manager had to decide whether or not to remove the inclined tower crane as well as the measures to employ. Due to the severity of the potential damage and urgency of the crisis situation, the project manager immediately looked for expertise and directions from the crisis management team to make an informed decision.

The senior manager in charge of construction technology from the Engineering department was first contacted. The communication between the site and command center continued expeditiously, constantly adding information about the extent of the damage, whether the tower crane should be removed immediately and how it would be removed. As the company maintained in-house resources and engineering capabilities, these additional resources and expertise were also involved and utilized in handling the critical engineering issues.

In line with greater needs from the project site, a special crisis task force comprising of management representatives and company in-house tower crane expert was assembled in the headquarters with the aim to visit the worksite and investigate the tower crane accident. Finally after the discussion and evaluation within the task force, it was decided to leave the tower crane intact as it was on site and take some necessary precautionary steps before further remedies were considered.

It could be seen that it was the emergent task, and not some structure that held the team members together. This project saw teams and individuals taking collective responsibilities for solutions to problems, other than resorting to defensive blame-laying behaviors. The result was a solution-building ethos and what became known on the project as a best-for-project culture. As the chief engineer from the Department of Engineering explained:

> We (our company) always have the preference to get the problem done first. Typically, the emerging problems were detected early and quickly dealt with. This is in contrast to behaviors where problems are often hidden in the hope that they will go away or that they will be dealt with by "somebody-else". Like this tower crane accident, we successfully and effectively setup a special crisis task force to solve the critical issues and take appropriate decisions.

Moreover, the focus on the problems also helped the crisis team members to hold back their level of anxiety when they were facing tricky issues. The attitude toward resolving problems by figuring out how to make things work out counterbalanced the project uncertainty and helped the members to overcome their fear and anxiety of dealing with uncertainties.

A supportive environment and organizational culture for solution-building that helped to maximize the contributions of the crisis management team and their members across disciplines and functional responsibilities was favorable. The culture of collective responsibility encouraged the kind of joint solution-building behaviors that this study had argued for. The management of the crisis was first of all focused on finding the solution, then on finding how and why it had occurred and lastly on reporting the incident.

The emergent establishment of the special crisis task force also described the organization's "optimal" response to crises. Although members might anticipate what kind of crises they might face, most companies could not accurately anticipate what would actually happen (Holder 2004). Organizations perceived and respond to crises through cultural structures that informed spontaneous behaviors and processes within the organizations, rather than through advance planning. As from Weick's (1995) perspective, it was not the crisis plan itself but rather the reflective process of planning that preceded it that enabled organizations to respond effectively to crises, regardless of specific "lists of do's and don't's or lessons learned" in the plan itself (Weick 1995, p. 52).

6.7.2.9 Shared Vision and Common Crisis Goal

As mentioned earlier, the company had embraced crisis management planning covering certain groups of site emergencies, aiming at "effective crisis response." The upper management encouraged and supported the deployment of the emergency procedures in all construction project worksites by performing regular inspections and internal audits, investigating near-miss accidents, reviewing safety performance at all levels, and communication of safety policies and evacuation procedures to site personnel. As project manager C explained,

> The existence of emergency planning and procedure is not enough. It cannot guarantee that the company is crisis-prepared. What sounds simple and probably effective on paper sometimes leads to ambiguity and conflict in the real word. As long as they are not under consideration, these concepts are nothing but words.
>
> We need to create a common mindset throughout the company and project worksite. For our project, we have conducted emergency response training and mock-drills on site. We have taken several measurements to improve safety and security in our daily operations. These included increasing the number of formal safety meetings with on-site staff, providing induction meetings for new staff at whatever level, and increasing site safety inspections. We also regularly conducted audits of the physical plant, equipment and conditions to identify potential emergencies.

It was argued that the common crisis goal should be co-created by all the project participants involved. Such a sense of purpose reinforced all the participants' willingness to give their time and energy to the activities demanded from the earthquake emergency, in order to help the process moved forward. The more the members interacted with each other, the more positive feedback was created, which reinforced the meaning of their collaborative activities. A vision without a supportive process and work environment was not sufficient to spur innovation and adaptive capability.

In the case of the tower crane accident, as the professionals and experts from the Department of Engineering in the company headquarter were actively and rapidly involved in solving the critical problem, their behavior shifted from being driven by self-interests toward a common pursuit of the company crisis vision. As Senge (1990) noticed, when shared goals and a commonality of purpose emerged, individuals do not sacrifice their personal interests to the larger team vision; rather, the shared vision became an extension of their personal visions. Such an extension of personal vision seemed to have occurred in the case study construction firm.

Developed through collaborative interactions among team members, the shared vision, which served as a shared mental model among the crisis management team, played a significant role in the team decision-making process. Numerous studies over the years have examined how groups or teams cooperated to make decisions and have identified processes and factors that differentiated successful decision-making groups from their less successful counterparts. Primarily, research showed that good teams monitored their performance and self-corrected; offered feedback; maintained awareness of roles and functions; communicated effectively, converged on a shared understanding of their situation and course of action; and coordinated their actions (Zsambok 1997).

The construction of shared understandings among the members of a group or an organization does not imply the absence of individual differences or friction. Choo (2001) noted that although shared meanings and purpose were constructed through sense-making, this does not mean that organizational members necessarily shared a common viewpoint. Weick (1995) also argued that conflict was an essential factor in human interactions. In fact, poly-vocality and encouragement of internal dissent could generate the kind of comprehensive and creative thinking that warded off crises to begin with (Tyler 2005).

This description sounded very much like the self-organizing behavior that underlied complex adaptive systems. For example, Cilliers (1998) identified several principles of self-organization in complex systems: these included a movement toward change and differentiation resulting from mutual feedback; cooperation among various agents; and reinforcement that eventually led the system to form coherent patterns.

6.7.2.10 Embracing Knowledge Ignorance and Adaptive Learning

In essence, the site project managers were entering a cosmology episode when the earthquake hit. They had never undergone such a crisis condition before. The cosmology episode, in Weick's (1993) terms, was described as the process of sense-making collapse, suggesting that organizations sought to make sense of crises, at least initially, by comparing them to previous events and past experiences. Weick (1993) illustrated the human reaction to cosmology episodes by such statements as "I have never been there before, I have no idea where I am, and I have no idea who can help me" (p. 634).

As explained by Weick (1998), human beings were constituted in such a way that we usually acted first and then try to rationalize the decision later. On many occasions, this rationalization occurred in conversation with others. Through the communication, we constructed well-thought-out accounts, and we became convinced of the rationality of our actions, which was from the start perhaps just an immediate response to our intuition. This process was called sense-making in the scientific context (Weick 1995). Hence, sense-making and communication were closely interconnected.

Despite the fact that the crisis managers were well aware of the significance of a sense-making approach, it was easy to rapidly fall into transmission thinking during the actual crises and to devote themselves solely to a one-way communication pattern. In certain aspects, one-way transmission was also necessary, for instance, distributing urgent warning messages. However, this was seldom sufficient as crisis communication was about dealing with paradoxes where there cannot be standardized formulas and procedures to follow.

This was especially true during the first 1 h of the massive earthquake occurring, when the conventional telecommunication systems including the telephone, mobile phone, and short-message services were severely interrupted. Reliable information was unavailable yet to develop an overall perspective of the existing situation at the site and what was happening elsewhere. The acute need for information to make sense of the current state created anxiety and uncertainty for both the project managers and the construction site workers to a particularly great extent. The uncertainty and ambiguity further affected the managers' capacity to communicate clearly about the earthquake crisis to the workers on site.

However, the site project managers had to learn quickly and take actions based on the best information at hand. This situation was quite similar to what Mitleton-Kelly (2005) described as living in a complex system that forced managers "to change their rules of interaction; to act on limited local knowledge, without knowledge what the systems as a whole is doing" (p. 27). Recognition and awareness of known knowledge, as well as knowledge ignorance (Harvey et al. 2001) or absence in crisis situations were vital in order to rapidly respond to the high levels of uncertainty for decision making. Managing this ignorance or lack of knowledge was therefore a kind of expertise or capability that helped managers deal with equivocal crisis information.

The project manager from project C described examples of how his project team responded quickly and made decisions during the very first few hours of the earthquake crisis occurring. One example was the utilization of more primitive communication resources to seek for more information when the primary communication channels broke down. For instance, the project team used radio to collect more accurate information of the earthquake in the first place, e.g., information relating to the earthquake magnitude, the location of the epicenter. Such emergent alternative communication resources facilitated the project manager to learn and adjust the crisis decisions to adapt to the current situation.

Another example was the adaptive reconfiguration of team skills. On the project C construction site, a senior worker who had experience with the rescue operations during the massive Tangshan earthquake of 1976 was the first person to sense and alert the earthquake event. He immediately informed the site supervisor and volunteered to offer assistance to rescue the stranded workers who were shocked by the sudden strong tremors from the elevated work platform. The spontaneous behavior rapidly drew attention to the project manager. Having understood the credibility in his work behavior, the project manager and site supervisor decided to engage the senior worker in the site emergency response team to assist with the rescue and evacuation procedures, and thus empowered him to provide them with the necessary and important earthquake related knowledge. Consequently, the coworkers demonstrated faith and trust in him and were willing to believe in him and to follow his guidance during the whole evacuation process.

In retrospect, the project manager emphasized that:

> The assignment of the experienced senior worker with prior earthquake experience as one of the core site emergency response team members had a very positive effect for the whole project team in response to the earthquake crisis. Of course, we seriously took into account his work credibility during the daily operations. Especially his confidence and sharing of knowledge and experience helped to alleviate the anxiety of his coworkers with less or no earthquake experience. I think how to allocate the resources available and to flexibly apply the crisis strategy on the spot according to the project characteristics and physical location is more important to deal with the crisis events. Merely stick to the crisis management plan cannot fully meet the crisis management requirements.

The site project team was able to base the decisions made on the knowledge at hand rather than waiting for more available information. Learning occurred when people openly confronted their knowledge of both knowing and ignorance. Organizations and individuals could gain knowledge by examining their ignorance (Stocking 1998). This actually reflected the adaptive learning capability and process of learning-by-doing which emphasized the role of action, self-reflection, and interaction. Such process followed the premise of a complex adaptive system that the microinteraction of highly localized exchanges between agents generated new patterns to exercise adaptability.

Learning-by-doing and self-reflection would create an environment where anxiety could be contained, trust could be developed, and issues could be resolved. Self-reflection was a capacity that needed to be learned at the individual, as well as the organizational levels. Since our beliefs influenced our perceptions of the world, we needed to examine our assumptions and beliefs to ensure that we do not hold internal contradictions. Similarly, an organization needed to examine the fundamental assumptions as the basis of the organization's vision, goals, and power structures and so on in order to assess whether these assumptions were aligned with the tacit understanding by organizational members of what the vision, goals, and power structures should be. By continuously identifying the contradictions, the organization was able to gain deeper understanding of issues at stake.

When highlighting the organizational environment that was conducive to learning, the project manager pointed out that creating an arena that supported learning had to be a continuous process within an organizational or team context. The arena could refer to physical areas in which organizational members might gather around and talk freely, and to a broader consideration that included providing time to engage them in nonproductive activities and the freedom to question organizational practices, policies, and power structures without inhibition (Rifkin and Fulop 1997). The manager's role was to help to create conditions that might open up the learning environment and to encourage learning by providing the arena for it to take place, instead of controlling and mandating the learning practice from a strict top-down approach.

Traditional management thinking gave central place to the command and control elements of leadership: defining the mission and steering the organization toward clearly defined goals that realized that mission. Clearly, complexity-based leadership discouraged these notions of control. Human beings affect, but not control, the development of the communicatively constituted human social system. They interacted with, rather than act on, others. Thus influence rather than control was the key characteristics of leadership in a learning-enabled organization. Managers could create conditions within the organization that facilitated and encouraged learning by providing the environment, physical or cultural, in which it could take place. The environment could be simply an organization climate where employees were free to talk about the organizational practices, policies, and power structure or to engage in nonproductive activities (Rifkin and Fulop 1997).

As explained by the deputy manager,

> We cannot rely on the crisis management plan too much. It cannot cover all the possible crises, plus, each crisis has its unique features. What is important is how we can improve our organizational learning and improvisation capability. We should encourage our employees to explore alternatives by speculating possible scenarios during the routine communication and discussions. By focusing on scenario building, e.g., what would have happened if and role playing, we might achieve more flexible crisis response.

This less-structured learning approach argued that learning was not necessarily a purposeful activity undertaken by an individual or group to meet goals set by upper management. It could be an ongoing experiential process through the interaction of organizational agents. The emerging phenomenon of learning in terms of dynamic patterns that resulted from interaction at the local level was compatible with some fundamental characteristics made by organizational theorists (Lewin and Regine 2003) who drew on complexity theory such as nonlinearity, emergence, and local interaction. As pointed out by Mitleton-Kelly (2003, p. 42), by drawing on complexity theory, the learning process is "an emergent property in the sense that it arises from the interaction of individuals and is not just the sum of existing ideas, but could well be something new and possibly unexpected. And once the ideas are articulated they form part of the history of each individuals and part of the shared history of the team."

It was commonly agreed that knowledge and learning were the very essential components of crisis communication management (Barton 2001; Coombs 2007; Fink 1986). However, the mainstream crisis literature placed the emphasis on preplanning the strategies and a set of tactics which the managers could rely on to make decisions without the need to build from scratch during the crisis, when they were confronted with either overwhelming information load or lack of information. The learning and sense-making process were mainly focused before and after a crisis to improve the organization's crisis management efficiency.

A complexity view of crisis events here extrapolated learning as an intrinsic part of the crisis management effort to adapt to the self-organizing nature of the complex system while the events were still unfolding. Learning, in this context, was an open-ended process that supported information and knowledge exchange with the goal of defining problems, as well as creating and experimenting different ways to solve problems. It was an interactive process that required much interaction between the people involved.

6.7.2.11 Nurturing Positive Behavior and Containing Anxiety

Crisis incidents were not only inconvenient times for organizations, but were also important psychological events experienced by individuals. It was usually the difficulties that people have in adapting to changes and uncertainties that created the behavioral and psychological problems (Applewhite 1965). While maximum attentions were desired to achieve effective crisis performance, however, this led to risk of the crisis managers generating excessive anxiety. Also the stress levels of employees were likely to be substantially exacerbated when they perceived the crisis situation as a threat to their health and safety. Such perceived threatening situations heightened employees' anxieties and uncertainty, as well as diminishing their sense of control over their own lives.

Moreover, high levels of stress might not only impair cognitive abilities, but also affected individual's emotional well-being (Kahn 1981), often leading to emotional distress. For examples, some of the workers were from the earthquake epicenter region. They were highly concerned about the safety and well-being of their families and those close to them. They were distracted by such concerns from attending to incoming crisis information. Individuals with serious emotional concerns or distractions were unable to concentrate on their jobs, which eventually adversely affected the organization.

On the one hand, it was argued that the power of each individual's perceptions of his or her situation during a crisis might ultimately determine his or her emotional and behavioral responses. According to Sillars and Parry (1982), the individuals' mental process might be influenced in various ways by high anxiety: "Several cognitive abilities which are instrumental to social perspective-taking may be affected, including hypothetical and abstract reasoning, information search, and the ability to differentiate and simultaneously consider multiple points of view" (p. 204). On the other hand, from the complexity point of view, the amount of anxiety and stress experienced for the employees also depended on the communication and interaction among their different perceptions of the crisis situation. A supportive environment fostering communication of the legitimacy of an individual's feelings and reactions was crucial to have positive effects on containing anxieties within the individual and team level.

As one of the project managers pointed out, in order to influence people, you have to show yourself to be an example of positive behavior. To become such an example, the project manager organized small group meetings with the jobsite workers to provide the necessary emotional and psychological support. He encouraged the employees to disclose their concerns and feelings with good humor and positive attitude. This created the opportunity for groups to meet to discuss the earthquake events in an emotionally safe way and to provide crisis-coping advice. An adaptive leader should be skilled at holding out the organizational or team members' level of anxiety, at least for a while, until the team was able to cope with it. However, if the leader held the members' level of anxiety for too long, members might become disengaged from the process and would not face their own problems. Thus, the leader would need to empower the team members to act and encourage them to take responsibilities for their own actions.

It was important to validate and share the employees' feelings and emotions because their primary function was to guide the behavior of individuals (Dillard 1998). It was important to adapt and respond to the environment and to make decisions about the appropriate ways to act in an environment (Scudder 1999). And at the same time, the management was also reassured by conscious recollection of the earthquake traumatic event.

Through interpersonal communication, the confidence and team moral were bolstered. Specifically, moral was the courage, discipline, confidence, enthusiasm, and willingness to endure hardship within a group (Guralnik 1980). Moral was the basic tenet by which organizations gauged their emotional selves. It was suggested that managers were often seen as effective if they have the ability to boost morale (Greenstone and Leviton 2002). Morale could be boosted or dismembered through communication (Williams 1978). Undermining the moral of the workers and the project team could influence the quality of work that in return affected the confrontational behavior of the participants.

Managing trust worthy and respectful relationship at a personal level was critical to facilitate this interpersonal communication. It was argued that organizational communication was about creating a community within the organization, which helped to define the identity of a group and to create a community spirit, which fit into organizational requirements (De Ridder 2004; Postmes et al. 2001). One critical factor that could influence feelings of belonging to a community within the organization was trust between the organizational members. The dominant perspective in the literature (Dirks and Ferrin 2001) was that trust resulted in distinctive effects such as more positive attitudes, higher levels of cooperation, and superior levels of performance (Jones and George 1998).

This was especially so in difficult times experienced by the workers exhibiting high stress. Trust could guide the actions of individuals and shape the perceptions of the coworkers and in this way guided the individual response to that action. The individuals usually exercised greater reliance on a trusted and credible peer to feel sufficiently psychologically safe as to explore their feelings and to reduce their anxieties and concerns.

Commitment and trust were clearly linked to the organizational climate and culture. Organizational climate was defined as the shared perceptions of organizational policies, practices, and procedures. At the level of the firm, positive attitudes could be encouraged by the organizational policy of the firm and the behavior and commitment of upper management. The issue of culture was seen to be driving behavior. If a firm has a strong culture that highlighted employee's needs and feelings as intrinsically important, then the sense of common purpose and cohesiveness was likely be sustained in times of crises.

Researchers have argued that employees' faith in an organization stemmed from a belief in its leaders' abilities to control and the commitment of upper management. The policy issue was of itself unlikely to shape the organizational culture, but a policy compiled with an active concern and involvement with managers to implement this policy was likely to realize an organizational culture of positive confrontational attitudes. In return, the crisis events were when organizations most needed close, trusting relationships with employees, as well as with other important external stakeholders. Indeed, failure to solicit employee input, argued by Stern (1987), could make the difference between keeping and losing key people.

In addition, to cope with the distractions of the employees, the company also devised a communication strategy tailored to address personal concerns of employees, e.g., gave a few days off to the workers to look after their family members who lived nearby the earthquake epicenter region. It was also suggested by the managers that approaches to address the employee's psychosocial needs could include incorporating stress reduction techniques into the organization's crisis management plan, training, and offering employees stress management classes or psychological counseling.

6.7.2.12 Final Remarks

The analysis of the Sichuan Tai Fa Construction Engineering Co., Ltd demonstrated how the organization and its crisis management team were able to create an integrated and collaborative communicative environment at both the company level and project jobsite level that supported its adaptation during the earthquake response stage. The crisis management team played a key role in sharing expertise and working together closely over time to develop shared crisis goals and mindset through cumulative, ongoing interaction. They favored collaborative dialog and problem solving that saw value in ambiguous and incomplete knowledge. By establishing trust and healthy interpersonal relationship, the employees' positive behaviors were mutually reinforcing, thus creating an environment supportive of cooperation. The influencing rather than control leadership helped to facilitate organizational learning and to contain the employees' anxiety to some extent.

Chapter 7
Conclusions and Recommendations

7.1 Summary

In an increasingly hostile business environment, organizations are susceptible to changes and uncertainties that can create a crisis situation at any stage of a project's life cycle. These crisis events can range from natural disasters causing damage to property and loss of life, accidents that may destroy a production plant and harm the environment, to unethical executives who make questionable decisions for monetary gains. Different organizations respond to their respective challenges with different adaptive capabilities that are influenced by diverse environmental and organizational variables, which affect their ability to learn and self-organize.

Complexity theory offers a new way of understanding how organizations function in complex and uncertain environments. As a general theory relating to the characteristics of nonlinear, self-organizing complex adaptive systems, complexity theory provides a useful analytical framework to explain the underlying patterns and process of communication and decision systems during a crisis. Considering the crisis response communication system as a complex adaptive system embedded within the organization's ecosystem implies that the organization's behavior is influenced by the interaction among the organization's subsystems and individual agents. Consequently, a critical role for adaptive and flexible communication management is to make sense of the complexity and uncertainty of the organizational environment and get insights on different organizational variables that may affect performance. By doing so, an organization is able to anticipate issues early and proactively develop strategies that facilitate adaptation.

This research provides a theoretical complexity-informed framework for effective crisis response communication management in a construction organization. Using this framework, organizational variables that can influence the control parameters of the crisis response communication system for the organization to creatively respond to the crisis were hypothesized. These organizational variables were derived from an extensive literature review on mainstream crisis management, organizational studies and management, and complexity theory. A questionnaire-based survey of general contractors in Chengdu, China who had experienced the

Z. Ying and L. Sui Pheng, *Project Communication Management in Complex Environments*, 163
DOI: 10.1007/978-981-4560-64-1_7, © Springer Science+Business Media Singapore 2014

massive Sichuan earthquake crisis in 2008 was undertaken. The survey was to determine whether the identified organizational variables were significant and to determine their level of importance in communication management in the aftermath of the earthquake crisis. Furthermore, for validation, two case studies were examined to offer an in-depth analysis of the emergent organizational communication structure, patterns of communication and behavior, and decision-making strategies of the Chengdu's construction firms in response to the earthquake crisis.

7.2 Conclusion on Research Hypotheses

This study tested the following major hypotheses:

- H1: The effectiveness and adaptivity of the communication management and decision-making practice of a construction organization in response to a crisis is influenced by organizational structure (OV1).
- H2: The effectiveness and adaptivity of the communication management and decision-making practice of a construction organization in response to a crisis is influenced by organizational culture (OV2).
- H3: The effectiveness and adaptivity of the communication management and decision-making practice of a construction organization in response to a crisis is influenced by information technology capability and information management system (OV3).
- H4: The effectiveness and adaptivity of the communication management and decision-making practice of a construction organization in response to a crisis is influenced by management and leadership style (OV4).
- H5: The effectiveness and adaptivity of the communication management and decision-making practice of a construction organization in response to a crisis is influenced by sense-making capability of the organization and members (OV5).
- H6: The effectiveness and adaptivity of the communication management and decision-making practice of a construction organization in response to a crisis is influenced by skill capability of members of the organization (OV6).

The survey results showed that all the six major hypotheses were supported to various degrees. Under H1, it was found that three out of the four organizational structure (OV1) related organizational variables, characterized by "a flat hierarchy structure (OV1.1)," "a highly bureaucratic top-down structure (OV1.2)," and "an ad hoc decentralized structure (OV1.3)" were supported. However, the organizational variable "a routine administrative structure (OV1.4)" was not supported. The findings indicated that the emergent organizational communication structure, which is sufficient to hold and exchange valid information throughout all parts of the organization and crisis management team, and through which information may flow easily and flexibly, has significant impact on the effectiveness and adaptivity of the crisis response communication management.

Under H2, all the eight organizational culture (OV2) related organizational variables were supported. This demonstrated that firms, which embrace adaptive learning and knowledge sharing, foster collaborative dialog and collective responsibility, and develop crisis common goal and a shared vision among the employees, are likely to have adaptive capability in the crisis response communication management.

Under H3, all the seven organizational variables related to the information technology capability and information management system (OV3) were supported. The findings indicated that the development of information technology and knowledge management is of paramount importance toward achieving effectiveness and adaptivity of the crisis response communication management.

It was found that twelve out of the twenty-one organizational variables related to management and leadership style (OV4) were supported. Therefore, H4 received partial support. In terms of communication management and process, "a flexible communication management process (OV4.1)," "a constructive conflict management (OV4.16)," "a defined sense of purpose (OV4.18)," and "crisis management plan (OV4.19)" was supported. Regarding the decision-making process, "a decentralized decision-making process (OV4.6)," "a delegation of decision authority (OV4.7)," "a proactive decision-making process (OV4.8)," "critical decisions are made by some dominant elites (OV4.10)," and "a long-term decision-making vision (OV4.15)" was supported. This indicated that top management should release certain level of decision-making authority and allow new processes to emerge freely within certain guidelines at local levels so as to adapt to the uncertainty and complexity of the crisis. And for leadership style, "an influencing, motivating and inspiring leadership (OV4.11)," "expertise to deal with crisis contingency (OV4.20)," and "ability to influence over others (OV4.21)" were also supported in this study. This is to be expected, as a leader by influencing rather than controlling the members and showing tolerance for ambiguity and uncertainty is more capable of holding in check the organizational anxiety in response to a crisis.

In this study, it was found that eleven out of the thirteen organizational variables related to the sense-making capability of organizations and members (OV5) under H5 were supported. Two organizational variables were rejected. The first was that "strong feeling of obligation and friendship (OV5.6)" is not significant for the effectiveness of crisis response communication. This can partly be explained that strong ties among the employees may lock the knowledge and information locally within the group and impede their dissemination to cross the groups for flexible communication. Another was that "defensive mechanism strategies to avoid anxiety (OV5.9)" is not important for crisis response communication. This may be because the defensive strategies would increase the organization's resistance to change and hinder its self-organization in response to an adaptive challenge.

Under H6, all the six organizational variables related to skill capabilities of members of the organization (OV6) were supported. The finding indicated that employees' skill capabilities and behavior significantly influence firms' adaptive capacity in communication management in response to a crisis.

7.3 Conclusion on Research Objectives

Figure 7.1 concludes that the research objectives stated in Sect. 1.4 have been achieved.

7.4 Major Contributions of the Study

This section presents major contributions of the study in three parts, i.e., major findings of the study, contributions to knowledge, and contributions to industry. The first section presents a brief description about the main findings of the survey and in-depth case-study analyses for crisis response communication management in Chengdu's construction firms in the aftermath of the Sichuan earthquake in 2008. A summary of the framework is illustrated in Fig. 7.2. The second section presents how the study is contributed to the body of knowledge about communication management in response to a crisis for construction firms. Finally, the third section sheds light on the study's contributions to the construction industry.

7.4.1 Major Findings of the Study

(a) In applying complexity theory to communication management in the context of crisis response, this research revealed that crisis response communication can be viewed as a complex adaptive system operating in an uncertain environment. Faced with environmental discontinuities, the organization self-organized at the edge of chaos, i.e., far from equilibrium, and created new communication structures that emerged out of the interaction between the interactions of diverse agents who came from different parts of the organization. Through the complexity lens, the interactive patterns and behaviors between individuals and organizations underlying the shifting social aggregations can be better understood.
(b) This research demonstrated that the complexity-informed framework can be successfully applied to study and analyze the communication patterns and decision-making practices of Chengdu's construction firms in response to the massive Sichuan earthquake in 2008. It can be argued that the proposed

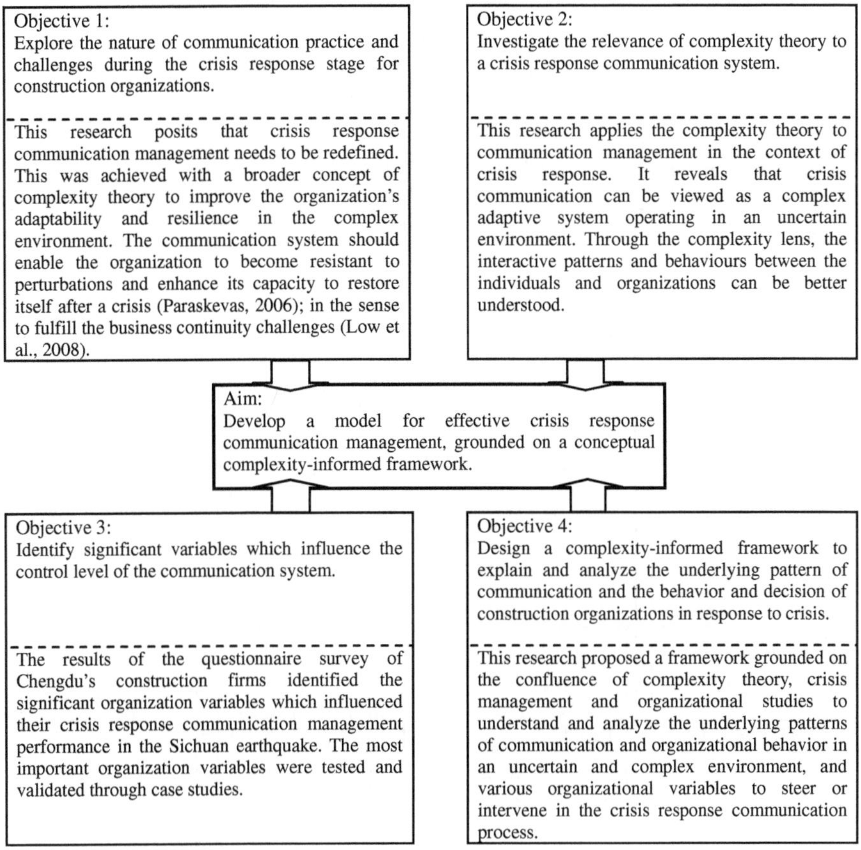

Fig. 7.1 Conclusion on research objectives

framework provides a useful platform to understand organizational behavior of firms in crisis events and various organizational variables to steer or intervene in the crisis response communication process.

(c) The study identified, through survey analyses and in-depth interviews with the professionals from Chengdu's construction firms who were directly involved in communication management of the building projects in response to the Sichuan earthquake in 2008, the important organizational variables which influence the control parameter of crisis response communication. A flat hierarchy structure, a delegation of decision authority style, developing crisis goal alignment, an influencing, motivating, and inspiring leadership style, learning environment, and transparency of intent and straightforward actions were considered to be the five top most important organizational variables for the Chengdu construction firms. The study also revealed that professionals from MES firms and LSE firms generally agreed on the significance of the identified organizational variables (Fig. 7.2).

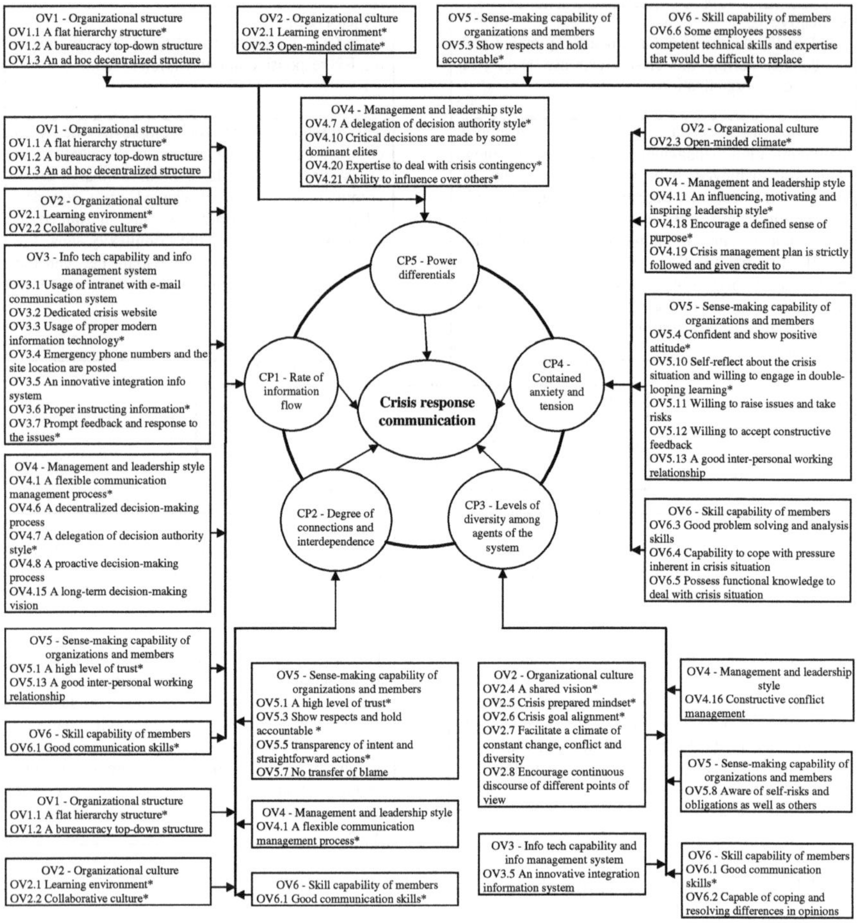

Fig. 7.2 Summary of framework

(d) The assessment of control parameters (i.e., the rate of information flow; agents' diversity; the connection and interdependency; the level of contained anxiety; and the degree of power differential) that was performed in the case studies supported the hypothesis that they need to reach critical points in order for an organization to creatively respond to the unforeseen events and overcome adaptive challenges. Therefore, the organizations need to progressively influence the control parameters toward the required control points by modifying their organizational variables and updating their strategies, in order to increase the adaptive capacity and flexibility of the organizations.

7.4.2 Contributions to Knowledge

The study contributes to the knowledge in crisis response communication management by developing a complexity-informed theoretical framework that expands the focus of conventional crisis management from uncertainty reduction, stable equilibrium, command-and-control to encompass uncertainty, far from equilibrium, multiple causes and outcomes, and the interaction within and between the organizations and their social contexts. The important methodological implication for studying and analyzing the crisis response communication is the relational view that draws an analogy between the crisis response communication and complex adaptive system, and the impact of the complexity based thinking on the way in which crises are identified and managed.

The proposed complexity-informed framework in this study does not tend to give organizations answers as to how problems should be solved or how communication can be made so that the targets can be achieved. Indeed, what should be bore in mind is that complexity theory does not provide recipes for success; it provides a new mental model to understand the organizational behavior in an uncertain and complex environment, and to get insights on the different organizational variables that may affect the performance. Here it is not argued that complexity-informed crisis management should completely replace traditional management approaches in all cases.

Another contribution is that this study proposed a complexity based approach to crisis response communication that tolerates the ambiguity and uncertainty, emphasizes by coping with the unexpected and uncertain, and encourages adaptive learning and sense making as a crisis evolves. The complexity based approach argues that effective crisis response communication management is not guaranteed by scientific planning and prescriptive decision making. It de-emphasizes the reliance on the prewritten procedures that may lead organizational members to ignore cues outside the prescribed framework. However, the complexity perspective encourages decision-makers not to abandon any hope of predicting but rather to rethink what we expect from that effort. Prediction is indeed possible within complex systems as long as we make a subtle but significant shift in our attitude, considering forecasting as "not the ability to foretell specific, well-defined events in space and time, but, at best, the ability to foretell the range of possible behaviors the system might adopt" (Van Uden et al 2001, p63).

The crisis response communication management performance can be influenced by different variables mutually influencing each other which results in limited predictability and control, and cannot be fully understood simply by analyzing individual variables. In the case study of the Chengdu construction firms, we have illustrated how the flexible communication structure, distributed intelligence, and collective responsibilities on CMT, shared crisis goal, collaborative dialog, adaptive learning and improvisation may complement precise planning as crisis management responds to the challenges brought by complexity's view as unstable, unpredictable, and intractable to control. The manager's ability to guide and influence an organization in a positive direction and behavior, and the diffused decision-making authority would also facilitate the organization to self-organize and adapt to the emergent change.

Lastly, this study examined how communication and information travel within a construction organization, and how individuals and teams acquire and transfer knowledge and information within a construction organization in the aftermath of a crisis. It bridged the gap in knowledge that the mainstream crisis communication literature generally emphasized on external communications with public relations or media relations, focusing on dissemination of information and monitoring audience reactions and concerns during the crisis (Barton 1993, 2001, Fearn-Banks 2007, Seeger Sellnow and Ulmer 2003). Crises are therefore the result of forces external to the organization that require a direct and identifiable response, and can be handled primarily by external communications. Within this study, particular attention has been drawn on how organizations find out about, make sense of, and make decisions for, and learn from rapidly changing situations. Hence it sees learning as an intrinsic part of the crisis management effort to adapt to the self-organizing nature of the complex system. The ability to recognize and open acknowledgment of knowledge ignorance helps the crisis managers deal with missing or equivocal information through adaptive learning. As a result, this helps to enhance the organization's adaptability and resilience in the event of a crisis and contribute to long-term business continuity management.

7.4.3 Contributions to the Construction Industry

(a) Considering the fact that crises in the construction industry are inevitable and not all crises cannot be foreseen, this study contributes to knowledge in the construction industry as the in-depth analyses presented could be used by the crisis managers and professionals to identify and evaluate significant variables to understand the organizational context and to steer the crisis response communication process in different crisis events.
(b) This study presents a new model of crisis response communication, grounded on the confluence of complexity theory, crisis management, and organizational studies. This new model would provide the professionals with a broader view on crisis management in which uncertainty, adaptiveness, and improvisation complement certainty, goal-orientation, and control. This would be helpful for the professionals to understand and analyze the underlying patterns of communication and organizational behavior in the situation of a crisis emerging. The model developed serves as a starting point/foundation for construction organizations to begin to reflect on their adaptive capacity and flexibility in managing crisis response communication by embracing complexity theory. As a result, these organizations can progressively influence the control parameters

toward the required control points by modifying their organizational variables and improving their strategies, in order to increase the adaptive capacity and flexibility of the organizations.

(c) The complexity-informed framework discussed in this study would assist the crisis managers and professionals to adopt the complexity-informed philosophy to develop adaptive leadership to facilitate learning and change process in a crisis in the construction industry. Complexity based leadership does not mean abdicating the traditional leadership functions; rather, it means redefining what the leader does. Adaptive leadership relates to the ability of an individual to influence a community to face its problems (Heifetz 1994). In this definition, the leader is not the one providing vision or solution to a problem. Rather, the leader is the one who is able to mobilize a community to define its own values for itself, enable it to identify the goals that support those values and implement the actions that need to be taken to achieve them. In practical terms, leaders are "not invested in stabling themselves as the ultimate authority;" instead of directing people, they "cultivate conditions where people could self-organize and restructure around the existing issues" (Lewin and Regine 2003, p173).

(d) This study would be useful for the construction industry to improve its adaptive capacity in the environment which is relatively uncontrollable and crisis-prone. One does not transform an entire industry or an organization overnight, but one can influence its members who will then become agents of change. Adaptive learning and improvisation skills are first developed at the individual level. As the individuals and professionals start interpreting the world from a complex adaptive system perspective, their mental models will change. A new mental framework will drive new behaviors, and over time, professionals will embed the new mental framework into their practices. Eventually these practices may be enacted within the organizations.

7.5 Limitation of the Research

The data for this study was collected from the Chengdu local construction firms who are registered with the Sichuan Construction Bureau under the General Building category. It may have excluded contractors who were also constructing building projects in the Chengdu metropolitan areas and who have also faced the earthquake crisis. This was because the scope of the research was narrowed to make it reasonably workable for an academic study, given the time and resource constraints.

The significance of the organizational variables was calculated based on the respondents' respective experience and perception of the level of importance of these variables, on a Likert scale. It is therefore unavoidable that different respondents may attach different values to the different points of the scale. However, efforts had been taken to address this issue, by conducting face-to-face interviews and discussion on the survey questions as mentioned in Chap. 6.

7.6 Future Recommendations

(a) This study presented in-depth analyses of crisis response communication management for construction firms in the event and aftermath of a massive earthquake crisis. Further works can be extended to other crisis events in the future.
(b) The proposed model based on complexity theory in this study can be further refined to gain a better understanding of the influence of the organizational variables on the control parameters of the crisis response communication process. A knowledge base of influence rules can be designed based on the impact of diverse organizational variables on the level of the control parameters. This knowledge-based system may provide more detailed information about the level of the control parameters and offer recommendations on how an organization should be managed to influence the control parameters to improve the flexibility and adaptability of its communication management in response to a crisis.
(c) Computational simulation models of crisis response communication within a construction organizational context can be developed that would facilitate the testing of hypothesis related to the influence of organizational variables on the control parameters. Using such computational models, the response behavior of the organization can be simulated and the impact of the organizational variables on the communication management performance can be assessed.

Appendix A
News on Promotion of Complexity Theory Study in Singapore Region

science.

THE STRAITS TIMES SATURDAY, FEBRUARY 14 2009 PAGE D10

New institute to study complexity theory

S'pore-based think-tank will look at global problems through multi-disciplinary work

By Liaw Wy-Cin

SINGAPORE has been identified as the Asian venue for a think-tank focusing on the cross-disciplinary study of global problems.

Work at the proposed research institute will be propelled by a new scientific field called the complexity theory.

This theory suggests that events and actions interact in complex ways to produce unpredictable effects. It has found applications in areas such as mathematics, computers, business, language and medicine.

It explains, for example, how poor lending decisions in the housing-loan mortgage sector in the United States triggered a global economic meltdown.

Two think-tanks which research issues using the theory are already up and running: One is the 25-year-old pioneering Santa Fe Institute in the US, and the other is the three-year-old Institute Para Limes in the Netherlands.

These two bodies, with Nanyang Technological University (NTU), are looking into opening the Asian version of the Santa Fe Institute here.

About 50 researchers from around the world gathered at NTU in the last two days for a conference on the scope of the theory, with speakers covering topics like Nobel Prize-winning discoveries, computers, economics and infectious diseases.

The conference, called Adaption, Order And Emergence, is a birthday tribute to a key contributor to the theory, Dr John Holland, who turned 80 last week.

He is credited as the founder of a field called genetic algorithms, a powerful computing tool that can generate the possible permutations of anything ranging from ways to build the best jet engine to ways of formulating the chemicals in a drug.

On how the complexity theory can be applied here, he said: "The more we know about complexity and how complex systems learn and adapt, the more we will know about innovation – and innovation is important for Singapore because you have no natural resources."

Guest of honour Peter Ho, who heads the civil service, cited the 2003 Sars crisis, last year's rice shortage and the current economic crisis as examples of "wild-card" or "black swan" events – high-impact events that appeared out of the blue and which were marked by their speed, surprise element and scope.

"They suggested to me that we are operating not in a linear world where cause and effect are clear, but in a more complex environment where cause and effect are difficult to discern, if at all," he said.

Mr Ho said it was not enough for Singapore to be a complex system.

"Ideally, we should be a complex, adaptive system. Maybe we will be one day," he said.

The Government's Risk Assessment and Horizon Scanning system – a computer tool which scans information from various sources to pick up potential threats, such as pandemic or terrorist ones, early – is one step in that direction, he said.

"It will help us to better anticipate strategic shocks that may be lurking just over the horizon," he said.

wycin@sph.com.sg

A better understanding of how complex systems work will help Singapore become more innovative, an important area due to its lack of natural resources, says Dr John Holland. PHOTO: NTU

(*Source* The Straits Time, Feb 14 2009)

Z. Ying and L. Sui Pheng, *Project Communication Management in Complex Environments*, DOI: 10.1007/978-981-4560-64-1, © Springer Science+Business Media Singapore 2014

Appendix B
Survey Questionnaire

Survey on Crisis Response Communication Management in Construction Firms

Note Please answer the questions based on your current construction firm.

Section A: General information of your firm and project

In this section, we wish to understand the background of your firm and information of your project which was affected by Sichuan earthquake. Please tick and/or answer the following questions:

1. Please indicate the major types of business and service your firm has provided: (You may tick more than one of the following)

☐ Civil engineering ☐ Property development ☐ Residential construction

☐ Office construction ☐ Factory construction ☐ Industrial building construction

☐ Others (please specify): ____________

2. Please indicate the type of your firm:

☐ Public listed firm ☐ Private Ltd firm ☐ Sole proprietorship/partnership

☐ Others (please specify): ____________

3. Total number of employees: ____________

4. Age of your firm: ____________

5. Please indicate the construction projects your firm has conducted, which was affected by the Sichuan earthquake crisis (Optional): ____________

6. Please indicate the location of the construction projects: ____________

7. Please specify the start and (expected) completion time of the construction projects:

Start time: (Year) ____________(Month)____________

(Expected) Completion time: (Year)____________(Month)____________

Z. Ying and L. Sui Pheng, *Project Communication Management in Complex Environments*, DOI: 10.1007/978-981-4560-64-1, © Springer Science+Business Media Singapore 2014

8. The contract value of the construction project: RMB¥ ____________

9. How do you consider the critical period in the aftermath of the earthquake to:

(a) Safety

☐ Within the first hour ☐ Within the first day ☐ Within the first week

☐ Within the first month ☐ Others (please specify): ____________

(b) Protect resources from pilfering and damages

☐ Within the first hour ☐ Within the first day ☐ Within the first week

☐ Within the first month ☐ Others (please specify): ____________

Section B: Organizational variables influencing crisis response communication management

In this section, we wish to understand how important the following variables are to your firm when managing communication in response to the crisis.

1. Please indicate the degree of significance of the following organizational structure in managing communication in response to crisis, based on your firm's experience (1 = least important, 2 = less important, 3 = important, good to have, 4 = more important, and 5 = most important):

Organizational structure						
1	A flat hierarchy structure	1	2	3	4	5
2	A highly bureaucracy top-down structure	1	2	3	4	5
3	An ad hoc decentralized structure	1	2	3	4	5
4	A routine administrative structure	1	2	3	4	5

2. Please indicate the degree of significance of the following organizational culture in managing communication in response to crisis, based on your firm's experience (1 = least important, 2 = less important, 3 = important, good to have, 4 = more important, and 5 = most important):

Organizational culture		
1	Learning environment encouraging constantly acquire and communicate knowledge and share critical crisis information	1 2 3 4 5
2	Collaborative culture fostering interaction and relationship between project participants	1 2 3 4 5
3	Open-mindedness climate encourage mutually trusting and honest interaction	1 2 3 4 5
4	Understand and commit to a shared vision. Such as employees are fully aware of different roles and responsibilities within their team/organization and willing to develop a shared understanding of the work-demand resulting from the earthquake crisis	1 2 3 4 5

(continued)

(continued)

Organizational culture	
5 Employees are fully aware of the emergency communication procedures (if applicable) and have crisis prepared mindset	1 2 3 4 5
6 Develop crisis goal alignment (e.g., to protect the resource from pilfering and damaging)	1 2 3 4 5
7 Facilitate a climate of constant change, conflict and diversity	1 2 3 4 5
8 Encourage continuous discourse of different points of view	1 2 3 4 5

3. Please indicate the degree of significance of the Information technology capability and information management system in managing communication in response to crisis, based on your firm's experience (1 = least important, 2 = less important, 3 = important, good to have, 4 = more important, and 5 = most important):

Information technology capability and information management system	
1 Usage of intranet within the company with e-mail communication system	1 2 3 4 5
2 A dedicated crisis or emergency website in the company's intranet system	1 2 3 4 5
3 Usage of proper modern information technology to facilitate communication	1 2 3 4 5
4 Emergency phone numbers and the site location are posted and clearly marked beside all site phones	1 2 3 4 5
5 An innovative integration information system is adopted. For example, capable of maintaining the IT system (IT data, financial archives, project documents, etc.) and restoring information or access backup system off site in case of any damages after earthquake	1 2 3 4 5
6 Employees have proper instructing information they need to know to protect themselves physically, e.g., how to evacuate dangerous construction area, find shelter for cover	1 2 3 4 5
7 Feedback and response to the issues are prompt in a timely fashion. For exapmple, quick contact with local public safety and emergency agencies to retrieve injured or stranded workers	1 2 3 4 5

4. Please indicate the degree of significance of management and leadership style in managing communication in response to crisis, based on your firm's experience (1 = least important, 2 = less important, 3 = important, good to have, 4 = more important, and 5 = most important):

Management and leadership style	
1 A flexible communication management process and procedures to allow for open communication	1 2 3 4 5
2 A rigid and centralized communication management process and procedures	1 2 3 4 5
3 Contractual rules and procedures are strictly followed during the response stage	1 2 3 4 5
4 Project managers or leaders are involved in great details in decision making	1 2 3 4 5
5 A participative decision-making process (democratic)	1 2 3 4 5

(continued)

(continued)

Management and leadership style		
6	A decentralized decision-making process	1 2 3 4 5
7	A delegation of decision authority style (empowerment), employees have amount of autonomy to make on-the-spot decisions during the crisis	1 2 3 4 5
8	A proactive decision-making process	1 2 3 4 5
9	Most decision are made through consensus	1 2 3 4 5
10	The critical decisions are made by some dominant elites who have the power to delegate decisions	1 2 3 4 5
11	A influencing, motivating, and inspiring leadership style	1 2 3 4 5
12	A control of authority leadership style (centralized)	1 2 3 4 5
13	A risk-taking preference leadership style (over-confident)	1 2 3 4 5
14	A habit-path resort preference leadership style (empiricism)	1 2 3 4 5
15	A long-term decision-making vision	1 2 3 4 5
16	A constructive conflict management for a well-managed conflict	1 2 3 4 5
17	A conflict avoidance and elimination preference	1 2 3 4 5
18	Encourage a defined sense of purpose	1 2 3 4 5
19	Crisis management/communication plan (if applicable) is strictly followed and given credit to	1 2 3 4 5
20	Manager's expertise to deal with crisis contingency	1 2 3 4 5
21	Manager's ability to Influence over others	1 2 3 4 5

5. Please indicate the degree of significance of sense-making capability of organizations and members in managing communication in response to crisis, based on your firm's experience (1 = least important, 2 = less important, 3 = important, good to have, 4 = more important, and 5 = most important):

Sense-making capability of organizations and members		
1	Project participants feel a high level of trust when communicating openly and truthful	1 2 3 4 5
2	Project participants offer help and provide reciprocal service,	1 2 3 4 5
3	Project participants show respects to one another, and hold accountable of each other	1 2 3 4 5
4	Project participants are confident and show a positive attitude to the project organization	1 2 3 4 5
5	Project participants demonstrate transparency of intent and straightforward actions	1 2 3 4 5
6	Project participants have a very strong feeling of obligation and friendship	1 2 3 4 5
7	Problems can be recognized at the very first sign and dealt with quickly, without transfer of blame to those who are not responsible	1 2 3 4 5
8	Our employees are fully aware of their risks and obligations as well as others	1 2 3 4 5
9	Develop defensive mechanism strategies to avoid anxiety	1 2 3 4 5
10	Our employees tend to self-reflection about the crisis situation and are willing to engage in double-loop learning	1 2 3 4 5
11	Our employees are willing to raise issues and take risks	1 2 3 4 5
12	Our employees are willing to accept constructive feedback	1 2 3 4 5
13	Our employees maintain a good interpersonal working relationship	1 2 3 4 5

6. Please indicate the degree of significance of skill capability of members of the organization in managing communication in response to crisis, based on your firm's experience (1 = least important, 2 = less important, 3 = important, good to have, 4 = more important, and 5 = most important):

Skill capability of members of the organization	
1 Our employees possess good communication skills to discourse and convey their opinions	1 2 3 4 5
2 Our employees is capability of coping and resolving differences in opinions	1 2 3 4 5
3 Our employees possess good problem solving and analysis skills to handle unforeseen circumstances. For exapmple, real-scene situational assessment, evaluate alternatives	1 2 3 4 5
4 Our employees possess capability to cope with pressure inherent in crisis situations	1 2 3 4 5
5 Our employees possess functional knowledge to deal with paradox and ambiguities inherent in crisis situations. For example, well trained with safety rules, emergency procedures, first aid	1 2 3 4 5
6 Some of our employees possess competent technical skills and expertise that would be difficult to replace	1 2 3 4 5

7. Please indicate below any other factors/variables that need to be considered in managing communication in response to crisis, based on your firm's experience (1 = least important, 2 = less important, 3 = important, good to have, 4 = more important, and 5 = most important):

Other variables to be considered in managing communication in response to crisis	
1	1 2 3 4 5
2	1 2 3 4 5
3	1 2 3 4 5
4	1 2 3 4 5
5	1 2 3 4 5

Section C: Information of respondent

1. Name of your firm: ____________
2. Your current position in your company: ____________
3. Number of years you have practiced in your profession: ____________

Thank you very much for taking the time to assist us in this study!

Name	Zhong Ying
Address	Department of Building School of Design and Environment National University of Singapore 4 Architectural Drive, Singapore 117566
Phone	(65) 8288 5369 (Mobile) (65) 6516 3513 (Office)
Email	yingzhong@nus.edu.sg

Appendix B
Survey Questionnaire (Chinese Version)

附件一:问卷调查大纲

第一部分 公司基本资料

本部分请您回答以下有关贵公司的问题:

1. **请问贵**公司主要提供以下那几**种**工程服**务**：（可以**选择**多**项**）

☐ 土木工程　　☐ 房地**产开发**　　☐ 民用住宅建**设**

☐ **办**公楼建**设**　　☐ 厂房**车间**建**设**　　☐ 工业区建**设**

☐ 其他 (**请说**明): ____________

Z. Ying and L. Sui Pheng, *Project Communication Management in Complex Environments*,
DOI: 10.1007/978-981-4560-64-1, © Springer Science+Business Media Singapore 2014

2. 请问贵公司的类型:

☐ 国营企业 ☐ 私营企业 ☐ 合资企业

☐ 其他 (请说明): ______________

3. 请问贵公司的总雇用员工数: ______________

4. 请问贵公司经营工程服务的年数: ______________

5. 请说明贵公司受到四川地震影响的工程项目: ______________

6. 请说明贵公司受到四川地震影响的工程项目的地址: ______________

7. 请指出贵公司受到四川地震影响的工程项目的开工时间和(预期)完工时间

开工时间: ______________ 年______________ 月

(预期)完工时间: ______________ 年______________ 月

8. 该受到四川地震影响的工程项目的合同价格: 人民币¥____________

9. 请问贵公司如何定义在地震应急反应中，对于处理以下事项的重要时期,

(a) 保护员工的生命安全

☐ 地震发生后的1小时内　☐ 地震发生后的1 天内　☐ 地震发生后的1星期内

☐ 地震发生后的1个月内　☐ 其它(请说明): ____________

(b) 保护在建项目的相关资源，防止偷窃和减小损害

☐ 地震发生后的1小时内　☐ 地震发生后的1 天内　☐ 地震发生后的1星期内

☐ 地震发生后的1个月内　☐ 其它(请说明): ____________

第二部分:影响危机应急反应阶段中沟通管理的组织变量

本部分请您回答关于贵公司在应急反应中的沟通管理中对以下因素的重视程度:

1. 请您根据贵公司的经验,评估并选择以下组织结构对于贵公司在危机应急反应阶段的沟通管理的重要程度 (1 = 非常不重要,2 = 比较不重要,3 = 一般重要,4 = 比较重要,5 = 非常重要):

组织结构						
1	扁平的层级结构	1	2	3	4	5
2	高度集权型结构	1	2	3	4	5
3	自主分权型结构	1	2	3	4	5
4	行政程序型结构	1	2	3	4	5

2. 请您根据贵公司的经验,评估并选择以下组织文化对于贵公司在危机应急反应阶段的沟通管理的重要程度(1 = 非常不重要,2 = 比较不重要,3 = 一般重要,4 = 比较重要,5 = 非常重要):

组织文化		
1	鼓励知识和重要危机信息共享的学习型组织文化	1 2 3 4 5
2	鼓励公司员工间的相互沟通交流的合作式组织文化	1 2 3 4 5
3	鼓励公司员工建立相互信任的开放式组织文化	1 2 3 4 5
4	明白如何从公司的视野考虑并且予以实施,例如员工充分了解项目或公司内部的不同责任和分工,对在公司应急反应中的工作需求达成共识。	1 2 3 4 5
5	公司员工充分了解公司应急措施的沟通程序和准则,有一定的危机准备意识	1 2 3 4 5

(continued)

(continued)

组织文化	
6 发展共同的危机应急管理目标,例如保护项目的相关资源,防止工地上的偷窃	1 2 3 4 5
7 促进不断变化,冲突和多样性并存的组织文化	1 2 3 4 5
8 鼓励不同意见和观点之间的沟通交流	1 2 3 4 5

3. 请您根据贵公司的经验,评估并选择以下信息技术能力和信息管理系统对于贵公司在危机应急反应阶段的沟通管理的重要程度(1 = 非常不重要,2 = 比较不重要,3 = 一般重要,4 = 比较重要,5 = 非常重要):

信息技术能力和信息管理系统	
1 使用公司内部互联网e-mail通信系统	1 2 3 4 5
2 在公司内部互联网内建立专有的地震危机管理网页	1 2 3 4 5
3 利用合适的现代信息通信技术进行沟通	1 2 3 4 5
4 建筑工地上有紧急联络电话的明显标示,和所处位置的标志	1 2 3 4 5
5 公司采取新型的综合信息管理体制,例如能够维持正常的信息管理 (项目文件,财政档案,等等);当受到地震破坏后,也能够恢复重要的项目信息数据或者使用项目工地以外的备用系统	1 2 3 4 5
6 公司员工能够接收到适当的指示信息,指引他们如何保护人生安全,例如如何从危险的工地位置撤出,如何寻求掩蔽场所,等等	1 2 3 4 5
7 公司在问题发生后的反馈和响应非常及时迅速,例如及时联络本地公共安全紧急管理部门,营救被困或受伤的员工	1 2 3 4 5

4. 请您根据贵公司的经验,评估并选择以下管理和领导模式对于贵公司在危机应急反应阶段的沟通管理的重要程度(1 = 非常不重要,2 = 比较不重要,3 = 一般重要,4 = 比较重要,5 = 非常重要):

管理和领导模式	
1 采取灵活的沟通管理程序,允许公开开放的交流	1 2 3 4 5
2 采取高度集权式的沟通管理程序	1 2 3 4 5
3 在应急反应阶段严格遵循合同的条例和程序	1 2 3 4 5
4 项目经理或高层领导参与决策过程各个细则	1 2 3 4 5
5 采取员工参与式的决策管理	1 2 3 4 5
6 采取分权式的决策管理	1 2 3 4 5
7 采取决策分权下授的管理模式,公司员工在处理紧急事件时有一定的自主性	1 2 3 4 5
8 大多数的决策是获得一致同意的	1 2 3 4 5
9 前瞻型主动的决策管理	1 2 3 4 5
10 公司应急沟通管理的关键决策由公司内部的某个(些)具备权利或影响力的人决定	1 2 3 4 5
11 影响,激励和鼓舞员工的领导风格	1 2 3 4 5
12 决策集权型的领导风格	1 2 3 4 5
13 风险性领导风格	1 2 3 4 5
14 倾向于经验主义的领导风格	1 2 3 4 5
15 倾向于有建设性的冲突管理体制	1 2 3 4 5

(continued)

(continued)

管理和领导模式	
16 倾向于冲突避免和减少的管理体制	1 2 3 4 5
17 采取详细明确的危机管理机制	1 2 3 4 5
18 鼓励目标明确	1 2 3 4 5
19 严格遵循公司的危机管理程序和沟通计划	1 2 3 4 5
20 领导者处理危机紧急事故的专业能力	1 2 3 4 5
21 领导者对于员工的影响力和感召力	1 2 3 4 5

5. 请您根据贵公司的经验,评估并选择以下组织和成员的危机意识感知能力对于贵公司在危机应急反应阶段的沟通管理的重要程度(1 = 非常不重要,2 = 比较不重要,3 = 一般重要,4 = 比较重要,5 = 非常重要):

组织和员工的危机意识感知能力	
1 员工有高度的信任感,公开和诚实的交流沟通	1 2 3 4 5
2 员工在沟通过程中有互利互惠的意识	1 2 3 4 5
3 员工相互表示尊重,有责任意识	1 2 3 4 5
4 员工对公司组织表示信任和乐观积极态度	1 2 3 4 5
5 员工在沟通中意图表达明确,直接了当	1 2 3 4 5
6 员工之间有很深厚的私人友谊和人情因素的考虑	1 2 3 4 5
7 问题可以在很初期征兆的时候被提出,并且迅速的解决,而不会被转移责任	1 2 3 4 5
8 员工不仅非常了解自己的任务风险,职责,也了解他人的任务职责	1 2 3 4 5
9 建立防御性的机制策略以避免焦虑	1 2 3 4 5
10 员工有自我反省的意识,愿意参与复杂性(双环)学习过程	1 2 3 4 5
11 员工愿意提出问题并承担风险	1 2 3 4 5
12 员工愿意接受建设性意见和建议	1 2 3 4 5
13 员工有很好的人际网络关系	1 2 3 4 5

6. 请您根据贵公司的经验,评估并选择以下组织成员的个人技能对于贵公司在危机应急反应阶段的沟通管理的重要程度(1 = 非常不重要,2 = 比较不重要,3 = 一般重要,4 = 比较重要,5 = 非常重要):

组织成员的个人技能	
1 员工拥有很好的交流沟通技能来表达自己的意见	1 2 3 4 5
2 员工有能力处理和解决存在不同意见和矛盾的情况	1 2 3 4 5
3 员工有足够的问题分析和解决能力来处理一些无法预料的事件,例如在地震应急环境中进行情况分析和评估,采取相应措施	1 2 3 4 5
4 员工有足够的能力应付危机环境下存在的各种压力	1 2 3 4 5
5 员工有足够的应急知识去应对地震环境下的不确定性,例如,员工曾接受专业的应急培训,安全培训,急救培训,等等	1 2 3 4 5
6 某些具备突出专业技术技能的员工在危机沟通管理中具有一定的影响力	1 2 3 4 5

7. 请您根据贵公司的经验,请指出其他一些在危机应急反应阶段的沟通管理中需要考虑的因素, 并请指出它们的重要程度(1 = 非常不重要,2 = 比较不重要,3 = 一般重要,4 = 比较重要,5 = 非常重要):

其他一些在危机应急反应阶段的沟通管理中需要考虑的因素					
	1	2	3	4	5
1					
	1	2	3	4	5
2					
	1	2	3	4	5
3					
	1	2	3	4	5
4					
	1	2	3	4	5
5					

第三部分:受访者信息

现在您已经填充完大部分的问题。我们非常感谢您付出了宝贵的时间和精力来帮助我们。您在这里所有的意见和答案都将受到严格保密。最后可否请您提供以下信息:

1. 贵公司名称: ____________
2. 您目前在贵公司的职位: ____________
3. 您已经从事您的职业多少年: ____________

非常感谢您协助我们的研究, 并祝您万事如意!

Appendix C
Interview Guide for Interview Questions

1. Please briefly describe your company background (e.g., number of employee, the major type of business).
2. Please describe the construction projects your company has conducted, which were affected by the Sichuan earthquake crisis in May 2008 (e.g., the contract value, the start and expected completion time).
3. What are the key communication challenges and issues during the crisis response stage? And what are the communication strategies and practices adopted by your company to overcome these challenges and issues?
4. How would you describe the information flow within your company during the crisis response stage?
5. Please describe typical patterns of communication within your company when the crisis hits.
6. Does your company have a formal crisis management or emergency plan to guide the communication actions and behavior? How was it implemented during the crisis? Did it guarantee a more effective response?
7. Did your company set up a dedicated crisis website for information sharing and for employees to monitor the overall situation? Were other media used to inform employees? Please describe the information through these other forms of media.
8. Did your company cultivate a crisis-awareness culture? Or did your employees share a common crisis prepared mindset? If yes, how do you think this culture influence the crisis management?
9. What leadership style and decision-making strategy do you think is needed to encourage the communication and interaction within the company?
10. How did your employees respond and react to the crisis? What capabilities and skills do you think are necessary for them to communicate and cope with pressure inherent in a crisis situation?
11. Can you describe more about the time in the midst of the Sichuan earthquake crisis? How did you contain the anxiety of your employee or yourself?

Z. Ying and L. Sui Pheng, *Project Communication Management in Complex Environments*,
DOI: 10.1007/978-981-4560-64-1, © Springer Science+Business Media Singapore 2014

12. What else about the crisis response communication experience would you like to share? And what are the practices or strategies you would suggest to improve the effectiveness of communication management in response to crisis?

Appendix C
Interview Guide for Interview Questions (Chinese Version)

附件二: 访问调查大纲

1. 请您简单说明一下贵公司的背景清情况(例如:总雇用员工数,公司的主要工程业务,等等)。
2. 请您简单描述一下贵公司受到在2008年5月份的四川大地震影响的工程项目情况(例如:项目的合同价格,项目的开工时间和(预期)完工时间)。
3. 请你根据贵公司的经验,指出在地震危机应急反应阶段所遇到的主要的沟通管理困难和挑战有哪些? 贵公司主要采用了那些沟通管理策略和办法来克服遇到的困难和挑战?
4. 请你描述一下在地震危机应急反应阶段,贵公司的信息流通情况。
5. 请您描述一下贵公司在地震危机应急反应阶段,采用的典型的沟通模式。.
6. 请问贵公司是否建立了一套正式的危机管理应急方案来引导沟通管理行为?如果有, 在危机应急反应阶段中,贵公司是如何执行的?您认为这样的危机管理应急方案能否保证作出有效的应急反应吗?
7. 请问贵公司是否建立了专门的地震危机管理网页,以用于信息共享和员工能够实时掌控信息? 公司员工是否有透过其他一些媒介来了解信息?如果有,请您进一步说明。
8. 请问贵公司是否蕴孕危机意识的文化氛围,或者是否贵公司的员工拥有共同的危机准备意思?如果有,请问您是如何看待这种公司文化对于危机管理的影响。
9. 请问您认为什么类型的领导模式和决策手段可以促进公司内部的交流沟通管理?
10. 请问贵公司员工对地震危机作出如何反应的?您认为员工需要什么样的沟通技能来应对危机反应阶段存在的压力?
11. 请您进一步描述在地震发生时的情况?请问您是如何抑制您的员工和您自己的焦虑的?您认为有哪些因素可以提高抑制焦虑的能力?
12. 请问你是否有其它一些在危机应急反应中的沟通管理经验和想法可以和我们分享?请问您能否就如何提高危机应急反应阶段的沟通管理,提出一些建议和意见?

非常感谢您协助我们的研究, 并祝您万事如意!

Z. Ying and L. Sui Pheng, *Project Communication Management in Complex Environments*,
DOI: 10.1007/978-981-4560-64-1,

References

Allen, M. W., & Caillouet, R. H. (1994). Legitimation endeavors: Impression management strategies used by an organization in crisis. *Communication Monographs, 61*, 44–62.
Allen, R. E. (1990). *The concise Oxford dictionary of current English*. Oxford: Clarendon Press.
Allen, P. M. (2001). A complex systems approach to learning in adaptive networks. *International Journal of Innovation Management, 5*(2), 149–180.
Allen, P. M., Cilliers, P., Midgley, G., & Richardson, K. (2003). *Complexity and its limits: The boundaries and ethics of understanding*. London: Routledge.
Anderson, P. (1999). Complexity theory and organizational science. *Organization Science, 10* (3), 216–232.
Anderson, P., Meyer, A., Eisenhardt, K., Carley, K., & Pettigrew, A. (1999). Introduction to the special issue: Applications of complexity theory to organization science. *Organization Science, 10*(3), 233–236.
Ankrah, N., & Langford, D. (2005). Architects and contractors: A comparative study of organizational cultures. *Construction Management and Economics, 23*(6), 595–607.
Applewhite, P. B. (1965). *Organizational behaviour*. Englewood Cliffs: Prentice Hall.
Arthur, W. B. (1994). Complexity in economic theory, inductive reasoning and bounded rationality. *American Economic Review, 84*(2), 406–411.
Arthur, W. B., Durlauf, S., & Lane, eds. (1997) *The Economy as an Evolving Complex System II*, Reading, Mass: Addison-Wesley.
Asian Disaster Reduction Center (2003) *Glossary on natural disaster*. Retrieved May 26, 2009, from www.adrc.or.jp.
Asian Disaster Reduction Center (2009) *Information on Disaster Risk Reduction of the Member Countries*. Retrieved June 26, 2009, from www.adrc.asia/nationinformation.php?NationCode=156&Lang=en&Mode=country.
Auyang, S. Y. (1998). *Foundations of complex system theories in economics, evolutionary biology, and statistical physics*. Cambridge: Cambridge University Press.
Axelrod, R. (1997). *The complexity of cooperation: Agent-based models of competition and collaboration*. Princeton: Princeton University Press.
Axley, S. R. (1996). *Communication at work: Management and the communication-intensive organization*. Westport: Quorum Books.
Baccarini, D. (1996). The concept of project complexity—a review. *International Journal of Project Management, 14*(4), 201–204.
Baguley, P. (1994). *Effective communication for modern business*. London: McGraw-Hill.
Baker, G., & Gollub, J. (1990). *Chaotic dynamics: An introduction*. New York: Cambridge University Press.
Bakhtin, M. M. (1981). *The dialogic imagination: Four essays*. Austin: University of Texas Press.

Z. Ying and L. Sui Pheng, *Project Communication Management in Complex Environments*, DOI: 10.1007/978-981-4560-64-1,

Barton, L. (1993). *Crisis in organizations: Managing and communicating in the heat of chaos*. Cincinnati: South-Western Publishing Co.

Barton, L. (2001). *Crisis in organizations II*. Cincinnati: South-Western.

Bavelas, A., & Barrett, D. (1951). An experimental approach to organizational communication. *Personnel, 27*(5), 366–371.

BBC News (2008). Retrieved February 27, 2009, from http://news.bbc.co.uk/2/hi/asia-pacific/7400614.stm.

Bendell, J. (2000). *Terms for endearment: Business NGOs and sustainable development*. Sheffield: Greenleaf Publishing.

Bennett, J. (1991). *International construction project management: General theory and practice*. Oxford: Butterworth-Heinemann.

Benoit, W. L. (1995). *Accounts, excuses, and apologies: A theory of image restoration strategies*. Albany: State University of New York Press.

Benoit, W. L. (1997). Image repair discourse and crisis communication. *Public Relations Review, 23*(2), 177–180.

Berelson, B. (1971). *Content analysis in communication research*. New York: Hafner Publishing Company.

Berg, B. L. (2006). *Qualitative research methods for the social sciences*. Boston: Allyn and Bacon Inc.

Bertelsen, S. (2003). *Complexity—Construction in a New Perspective*. Presented at the IGLC-11 Annual Conference, Blacksburg, August 2003. Retrieved August 12, 2007, from www.bertelsen.org/.../pdf/Complexity%20%20Construction%20in%20a%20New%20Perspective.pdf.

Birch, J. (1994). New factors in crisis planning and response. *Public Relations Quarterly, 39*, 31–34.

Blalock, H. M. (1989). *Power and conflict: Toward a general theory*. Newbury Park: Sage.

Boddy, D., & Buchanan, D. (1999). *Take the lead: Interpersonal skills for project managers*. New York: Prentice-Hall.

Bohm, D. (1985). *Unfolding meaning: A weekend of dialogue with David Bohm*. New York: Ark Books.

Boisot, M., & Child, J. (1999). Organizations as adaptive systems in complex environments: The case of China. *Organization Science, 10*(3), 237–252.

Boje, D. M. (2000). Phenomenal complexity theory and change at Disney: Response to letiche. *Journal of Organizational Change Management, 13*(6), 456–475.

Booth, S. A. (1993). *Crisis management strategy—competition and change in modern enterprises*. London: Routledge.

Botkin, D. (1990). *Discordant harmonies: A new ecology for the twenty-first century*. New York: Oxford University Press.

Boyd, D., & Wild, A. (2003). The tavistock studies into the building industry: Communications in the building industry (1965) and interdependence and uncertainty (1996). In Murray, Mike & Langford, David (Eds.), *Construction Reports* (pp. 1944–1998). Oxford: Blackwell Science.

Briner, W., Hastings, C., & Geddes, M. (1996). *Project leadership*. Aldershot: Gower.

Brodbeck, P. W. (2002). Complexity theory and organization procedure design. *Business Process Management Journal, 8*(4), 377–402.

Brockner, J. (2008). Toward an understanding of when executives see crisis as opportunity. *The Journal of Applied Behavioral Science, 44*(1), 94–115.

Bronn, P. S., & Olson, E. L. (1999). Mapping the strategic thinking of public relations managers in a crisis situation: An illustrative example using conjoint analysis. *Public Relation Review, 25*(3), 351–368.

Burawoy, W. (1991). *Unbound: Power and resistance in the modern metropolis*. California: University of California Press.

Burke, W. W. (1986). Leadership as empowering others. In S. Srivastva (Ed.), *ExecutivePower*. San Francisco: Jossey-Bass.

Burton, R. M., & Obel, B. (2004). *Strategic organizational diagnosis and design: The dynamic of fit* (3rd ed.). New York: Springer.

Business Dictionary (BD) (2009). Retrieved March 3, 2009, from http://www.businessdictionary.com/definition/organizational-structure.html.

Byrne, D. (1998). *Complexity theory and the social sciences: An introduction*. London: Routledge.

Caponigro, J. R. (2000). *The crisis counselor: A step-by-step guide to managing a business crisis*. Lincolnwood: Contemporary Books.

Capra, F. (2002). *The Hidden connections: Integrating the biological, cognitive and social dimensions of life into a science of sustainability*. New York: Doubleday.

Capra, F. (1996). *The web of life*. London: HarperCollins.

Carlisle, Y., & McMillan, E. (2002). Thinking differently about strategy: Comparing paradigms. *Proceedings of the 16th Annual Conference of the Australian and New Zealand Academy of Mgtt*, Australia. Retrieved July 10, 2008, from http://oro.open.ac.uk/7499/1/Thinking_Differently_Final.pdf.

Carroll, T., & Burton, R. M. (2000). Organizations and complexity: Searching for the edge of chaos. *Computational and Mathematical Organization Theory, 6*, 319–337.

Caywood, C., & Stocker, K. P. (1993). The ultimate crisis plan. In J. Gottschalk (Ed.), *Crisis-response: Inside stories on managing image under siege* (pp. 409–428). Detroit: Gale Research.

Chen, J. H. (1998). The characteristics and current status of China's construction industry. *Construction Management and Economics, 16*(6), 711–719.

Chia, R. (2002). Essay: Time, duration and simultaneity: Rethinking process and change in organizational analysis. *Organization Studies, 23*(6), 863–868.

Choo, C. W. (2001). The knowing organization as learning organization. *Education + Training, 43*(4–5), 197–205.

Church, A. H. (1996). Giving your organizational communication C-P-R. *Leadership and Organization Development Journal, 17*(7), 4–11.

Cicmil, S. (2003). *Knowledge, interaction and project work: The perspective of complex responsive processes of relating*. Presented at the 19th EGOS Colloquium, Project Organizations, Embeddedness and Repositories of Knowledge. Copenhagen, 3–5 July 2003.

Cicmil, S. (2005). Participation, reflection, and learning in project environments—a multiple perspective agenda. In P. Love, Z. Irani, & S. W. Fong (Eds.), *Managing knowledge in project environments* (pp. 155–179). Oxford: Butterworth-Heinemann.

Cilliers, P. (1998). *Complexity and Post Modernism: Understanding Complex Systems*. New York: Routledge.

Comfort, L. K. (1994). Self organization in complex systems. *Journal of Public Administration Research and Theory, 4*(3), 393–410.

Comfort, L. K. (1996). *Self-organization in Disaster Response: The Great Hanshin*, Japan Earthquake of January 17, 1995, Quick Response Report No. 78, Natural Hazards Center, University of Colorado, Boulder. Retrieved June 15, 2008, from www.colorado.edu/IBS/hazards/qr/qr78.html.

Comfort, L. K. (1999). *Shared risk: Complex systems in seismic response*. Oxford: Pergamon Press.

Comfort, L. K., Sungu, Y., Johnson, D., & Dunn, M. (2001). Complex systems in crisis: Anticipation and resilience in dynamic environments. *Journal of Contingencies and Crisis Management, 9*(3), 144–158.

Cooke-Davies, T., Cicmil, S., Crawford, L., & Richardson, K. (2008). We're not in Kansas anymore, Toto: Mapping the strange landscape of complexity theory, and its relationship to project management. *Engineering Management Review IEEE, 36*(2), 5–21.

Coombs, W. T. (1995). Choosing the right words: The development of guidelines for the selection of the "appropriate" crisis response strategies. *Management Communication quarterly, 8*, 447–476.

Coombs, W. T. (2007). *Ongoing crisis communication: Planning, managing and responding* (2nd ed.). Thousand Oaks: Sage.

Crane, A., & Livesey, S. (2003). Are you talking to me? Stakeholder communication and the risks and rewards of dialogue. In J. Andriof, S. Waddock, B. Husted, & S. Sutherland Rahman (Eds.), *Unfolding Stakeholder Thinking 2: Relationships, Communication, Reporting and Performance* (pp. 39–52). Sheffield: Greenleaf Books.

Crutchfield, J., Farmer, D., Pachard, N., & Shaw, R. (1986). Chaos. *Scientific American, 255*, 46–57.

CSAO (Construction Safety Association of Ontario). (2003). *Emergency response planning for construction projects*. Retrieved June 5, 2009, from www.csao.org/images/pfiles/278_EmergencyResponsePlanning.

Dainty, A., Moore, D., & Murray, M. (2006). *Communication in construction: Theory and practice*. London: Taylor and Francis Group.

Deetz, S. (1995). Transforming communication, transforming business: Stimulating value negotiation for more responsive and responsible workplaces. *International Journal of Value-Based Management, 8*(3), 255–278.

De Ridder, J. (2004). Organisational communication and supportive employees. *Human Resource Management Journal, 14*(3), 20–30.

Dillard, J. P. (1998). The role of affect in communication, biology, and social relationships. In P. A. Andersen & L. K. Guerrero (Eds.), *Handbook of communication and emotion: Research, theory, application and contexts, xvii–xxxii*. San Diego: Academic Press.

Dirks, K. T., & Ferrin, D. L. (2001). The role of trust in organisational setting. *Organisational Science, 12*(4), 450–467.

Dorner, D. (1996). *The logic of failure: Recognizing and avoiding error in complex situation*. Reading: Perseus.

Eisenberg, E. M., & Goodall, H. L, Jr. (1993). *Organizational communication: Balancing creativity and constraint*. New York: St Martins Press.

Emmitt, S., & Gorse, C. A. (2003). *Construction communication*. Oxford: Blackwell.

Englehardt, C. S., & Simmons, P. R. (2002). Organizational flexibility for a changing world. *Leadership and Organizational Development Journal, 23*(2), 113–121.

Espejo, R. (2006). Reflections on power, democracy and communications. *Cybernetics and Human Knowing, 13*(3–4), 144–152.

Feagin, J. R., Orum, A. M., & Sjoberg, G. (1991). *A case for the case study*. Chapel Hill: University of North Carolina Press.

Fearn-Banks, K. (2007). *Crisis communications: A casebook approach* (3rd ed.). Mahwah: Erlbaum.

Feldberg, M. (1975). *Organization behavior: Text and cases*. Cape Town: Juta and Company.

Fellows, R., & Seymour, D. E. (eds) (2002). *Perspectives on Culture in Construction*. CIB report, 275.

Fink, S. (1986). *Crises management: Planning for the inevitable*. New York: AMACOM.

Fink, S. B., Beak, J., & Taddeo, K. (1971). Organizational crisis and change. *Journal of Applied Behavioral Science, 7*, 15–37.

Fiske, J. (1990). *Introduction to communication studies*. London: Routledge.

Fitzgerald, L. A. (1996). *Organizations and other things fractal: A primer in chaos for agents of change*. Denver: The Consultancy Inc.

Flower, J. (1993). The power of chaos. *Healthcare Forum Journal, 36*(5), 48–55.

Fonseca, J. M. L. (1998). *Innovation: A property of complex adaptive social systems*, unpublished thesis, University of Hertfordshire.

Fontana, A., & Frey, J. H. (1994). Interviewing: The art of science.In N. K. Denzin & Y. S. Linconln (Eds.), *Handbook of Qualitative Research*. Thousand Oaks: Sage Publication, Inc.

Foster, D., & Jonker, J. (2005). Stakeholder relationships: The dialogue of engagement. *Corporate Governance, 5*(5), 51–57.

Fraser, L. (2009). *Human error—study projct by Lynn Fraser*. Retrieved July 6, 2009, from www.chemeng.ed.ac.uk/~jskillin/teaching/.../long**humanerror**.pdf.
Freeman, R. E. (1984). *Strategic management: A stakeholder approach*. Boston: Pitman.
Frooman, J. (1999). Stakeholder influence strategies. *Academy of Management Review, 24*, 191–205.
Galbraith, J. R. (1973). *Designing complex organizations*. New York: Addison-Wesley.
Gell-Mann, M. (1994). Complex adaptive systems. In A. C. George, P. David, & M. David (Eds.), *Complexity: Metaphors. Models and Reality* (pp. 17–29). Reading: Addison-Wesley
Griffin, D. (2002). *The emergence of leadership, linking self-organization and ethics*. London: Routledge.
Gigliotti, R., & Ronald, J. (1991). *Emergency planning for maximum protection*. New York: Butterworth-Heinemann.
Gilchrist, A. (2000). The well-connected community: networking to the edge of chaos. *Community Development Journal*, 35, 264–275.
Gill, J., & Johnson, P. (1997). *Research methods for manager* (2nd ed.). London: Paul Chapman Publishing Ltd.
Gilpin, D. R., & Murphy, P. J. (2008). *Crisis management in a complex world*. New York: Oxford University Press.
Gleick, J. (1987). *Chaos: The making of a new science*. New York: Viking.
Goldberg, J., & Markoczy, L. (2000). Complex rhetoric and simple games. *Emergence, 2*(1), 72–100.
Goldspink, C. (2000). Contrasting linear and non-linear perspectives in contemporary social research: Organization theory. *Emergence, 2*(2), 72–101.
Greenstone, J. L., & Leviton, S. C. (2002). *Elements of crisis intervention: Crises and how to respond to them* (2nd ed.). Pacific Grove: Brooks/Cole.
Greiner, L. E., & Schein, V. E. (1988). *Power and organization development*. New York: Addison-Wesley.
Griffin, D. (2002). *The emergence of leadership, linking self-organization and ethics*. London: Routledge.
Grunig, J. E. (1992). *Excellence in public relations and communication management*. Hillside: Lawrence Erlbaum.
Grunig, J. E., & White, J. (1992). The effect of worldview on public relations theory and practice. In J. E. Grunig (Ed.), *Excellence in public relations and communication management* (pp. 31–64). Hillsdale: Lawrence Erlbaum.
Guralnik, D. B. (Ed.). (1980). *Webster's new world dictionary of the American language* (2nd ed.). New York: Prentice-Hall Press.
Haigh R., Amaratunga, D., & Kerimanginaye, K. (2006). An exploration of the construction industry's role in disaster preparedness, response and recovery. *In the* proceedings of *the Annual Research Conference of the Royal Institution of Chartered Surveyors*, held at University College London. Retrieved June 15, 2009, from http://www.preventionweb.net/files/9086_COB06Haigh1.pdf.
Hale, E. J., Dulek, R. E., & Hale, D. P. (2005). Crisis response communication challenges—Building theory from qualitative data. *Journal of Business Communication, 42*(2), 112–134.
Hallgren, M., & Wilson, T. L. (2008). The nature and management of crises in construction projects. *International Journal of Project Management, 26*(8), 830–838.
Hall, R. M. (1991). *Organization, Structure, Process, Outcomes*. Englewood Cliffs: Prentice Hall.
Handy, C. (1994). *The empty coat*. UK: Arrow.
Harrald, J. R. (2006). Agility and discipline: Critical success factors for disaster response. *The Annuals of The American Academy*, Retrieved June 1, 2009, from www.bus.lsu.edu/centers/sdmi/files/Annals.pdf.
Harrison, J. S., & St John, C. H. (1996). Managing and partnering with external stakeholders. *Academy of Management Executive, 10*(2), 46–51.

Harvey, M. G., Novicevic, M. M., Buckly, M. R., & Ferris, G. R. (2001). A historic perspective on organizational ignorance. *Journal of Management Psychology, 16*(6), 449–468.

Hatch, M., & Tsoukas, H. (1997). *Complex thinking about organizational complexity: the appeal of a narrative approach to complexity theory*. Paper presented at the American Academy of Management Meeting, Boston, August

Hayek, F. A. (1954). The use of knowledge in society. *American Economic Review, 35*(4), 519–530.

Hayles, K. (1990). *Chaos bound*. Ithaca: Cornell University Press.

Heath, R. (1998). Working under pressure: Crisis management, pressure groups and the media. *Safety Science, 30*, 209–221.

Health, R. L., Pearce, W. B., Shotter, J., et al. (2006). The processes of dialogue: Participation and legitimation. *Management Communication Quarterly, 19*(3), 341–75.

Hearit, K. M., & Courtright, J. L. (2004). A symbolic approach to crisis management: Sears' defense of its auto repair policies. In D. P. Millar & R. L. Heath (Eds.), *Responding to crisis: A rhetorical approach to crisis communication* (pp. 201–212). Mahwah: Erlbaum.

Heifetz, R. A. (1994). *Leadership without easy answers*. Cambridge: The Belknap Press of Harvard University Press.

Hermann, C. F. (1963). Some consequences of crises which limit the viability of organizations. *Administrative Science Quarterly, 8*, 61–82.

Heylighen, F., Cilliers, P., & Gershenson, C. (2006). Complexity and Philosophy. Retrieved March 15, 2008, from http://cogprints.org/4847/1/ComplexityPhilosophy.doc.pdf.

Hinnings, C. R., Hickson, D. J., Pennings, J. M., & Schneck, R. E. (1974). Structural conditions of intraorganizational power. *Administrative Science Quarterly, 16*, 22–44.

Holder, T. L. (2004). Construction response during uncertainty: Organizing for crisis. In D. P. Millar & R. L. Heath (Eds.), *Responding to crisis: A rhetorical approach to crisis communication* (pp. 51–62). Mahwah: Erlbaum.

Holland, J. (1995). *Hidden order: How adaptation builds complexity*. London: Addison-Wesley.

Holland, J. (1998). *Emergence: From chaos to order*. USA: Oxford University Press.

Holtzhausen, D. R. (2000). Postmodern values in public relations. *Journal of Public Relations Research, 12*, 93–114.

Holtzhausen, D. R. (2002). Towards a postmodern research agenda for public relations. *Public Relations Review, 28*(3), 251–264.

Horgan, J. (1996). *The end of science: Facing the limits of knowledge in twilight of the scientific age*. Reading: Addison-Wesley.

Horsley, J. S., & Barker, R. T. (2002). Toward a synthesis model for crisis communication in the public sector. *Journal of Business and Technical Communication, 16*(4), 406–441.

Ireland, V. (1985). The role of managerial actions in cost, time and quality performance of high-rise building projects. *Construction Management and Economics, 3*(1), 59–87.

Irvine, R. B., & Millar, D. P. (1996). Debunking the stereotypes of crisis management: The nature of business crises in the 1990s. In L. Barton (Ed.), *New avenues in risk and crisis management* (Vol. 5, pp. 51–63). Las Vegas: University of Nevada Las Vegas Small Business Development Center.

Jackson, M. C. (2000). *Systems approaches to management*. New York: Kluwer.

James, E. H. (2010). Crisis leadership. *SSRN*. Retrieved June 22, 2010, from http://papers.ssrn.com/sol3/papers.cfm?abstract_id=1281843&rec=1&srcabs=224055.

James, E. H., & Lynn, P. W. (2005). Leadership as (un)usual: How to display competence in times of crisis. *Organizational Dynamics, 34*(2), 141–152.

Jarman, A., & Kouzmin, A. (1990). Decision pathways from crisis—a contingency theory simulation heuristic for the challenger space disaster (1983–1988). In A. Block (Ed.), *Contemporary crisis—law, crime and social policy* (pp. 399–433). Netherlands: Kluwer Academic Press.

Jarvenpaa, S. L., & Ives, B. (1994). The global network organization of the future: Information management opportunities and challenges. *Journal of Management Information Systems, 10* (4), 25–57.

Jick, T. D. (1979). Mixing qualitative and quantitative methods: Triangulation in action. In J. van Mannen (Ed.), *Qualitative methodology* (pp. 135–148). London: Sage Publisher.

Jones, G., & George, J. (1998). The experience and evolution of trust: Implications for cooperation and teamwork. *Academy of Management Review, 23*(3), 531–546.

Kahn, R. L. (1981). *Work and health*. New York: Wiley.

Kaklauskas, A., Zavadskas, E. K., Bagdonavicius, A., Kelpsiene, L., Bardauskiene, D., & Kutut, V. (2010). Conceptual modelling of construction and real estate crisis with emphasis on comparative qualitative aspects description. *Transformations in Business and Economics*, *9*(1 (19)), 42–61.

Kapucu, N., & Van Wart, M. (2006). The emerging role of the public sector in managing extreme events: Lessons learned. *Administration and Society, 38*(3), 279–308.

Kauffman, S. A. (1993). *Origins of order: Self-organization and selection in evolution*. Oxford: Oxford University Pess.

Kauffman, S. A. (1996). *At Home in the universe: The search for laws of self-organization and complexity*. London: Penguin Books.

Kauffman, S. A. (2000). *Investigations*. New York: Oxford University Press.

Kaufman, D. (2006). *Complexity and communications: A conceptual framework*. Prepared for the Annual Meeting of the International Studies Association. Retrieved July 2008, from http://www.allacademic.com//meta/p_mla_apa_research_citation/1/0/0/4/4/pages100444/p100444-1.php.

Kelly, S. (1999). *The complexity advantage: How the science of complexity can help your business achieve peak performance*. New York: Business Week Books.

Kelly, S., & Allison, M. A. (1999). *The complexity advantage*. New York: McGraw-Hill.

Kometa, S. T. (1995). *An evaluation of client-generated risks in construction consultancies*. Unpublished doctoral thesis, University of Wolverhampton, England.

Krippendorff, K. (1989). On the ethics of constructing communication. In Brenda Dervin, Larry Grossberg, Barbara J. O'Keefe, & Ellen Wartella (Eds.), *Rethinking communication: Paradigm issues* (Vol. I). Newbury Park: Sage Publications.

Krugman (1996) *The Self Organizing Economy*. Oxford: Blackwell.

Kuhn, T. S. (1996). *The structure of scientific revolutions* (3rd ed.). Chicago: The University of Chicago Press.

Langton, C. G. (1989). *Artificial life*. Redwood City: Addison-Wesley.

Langton, C. G., Taylor, C., Farmer, J. D., & Rasmussen, S. (Eds.). (1992). *Artificial Life: Santa Fe Institute Studies in the Sciences of Complexity* (Proceedings Vol. 10). Redwood City: Addison-Wesley.

Lawrence, P. C., & Lorsch, J. W. (1967). *Organization and environment: Managing differentiation and integration*. Boston: Graduate School of Business Administration, Harvard University.

Leavitt, H. J. (1951). Some effects of certain communication patterns on group performance. *Abnormal and Social Psychology, 46*(1), 38–50.

Lerbinger, O. (1997). *The crisis manager facing risk and responsibility*. Mahwah: Lawrence Erlbaum.

Lesser, R., & Prusak, L. (2004). *Creating value with knowledge: Insights from the IBM institute for business value*. Oxford: Oxford University Press.

Levinthal, D. A., & Warglien, M. (1999). Landscape design: Designing for local action in complex worlds. *Organization Science, 10*(3), 342–357.

Lewin, A. Y. (1999). Application of complexity to organizational science. *Organization Science, 10*(3), 215.

Lewin, R. (1993). *Complexity: Life at the edge of chaos*. London: Phoenix.

Lewin, R., & Regine, B. (1999). *The Soul at Work*. London: Orion Business.

Lewin, R., & Regine, B. (2001). An organic approach to management. *The Cap Gemini Ernst and Young Center for Business Innovation, Perspective on Business Innovation Journal, 4*, 19–26.

Lewin, R., & Regine, B. (2003). The core of adaptive organization. In E. Mitleton-Kelly (Ed.), *Complex systems and evolutionary perspectives on organizations: The application of complexity theory to organizations* (pp. 167–183). New York: Pergamon Press.

Levin, S. A. (1998). Ecosystems and the biosphere as complex adaptive systems. *Ecosystems, 1*, 431–436.

Liedtka, J. (1998). Constructing an ethic for business practice: Competing effectively and doing good. *Business and Society, 37*(3), 254–280.

Lincoln, Y. S., & Guba, E. G. (1985). *Naturalistic inquiry*. Galifarnia: Sage Publications.

Lindlof, T. R. (1995). *Qualitative communication research methods*. Thousand Oaks: Sage Publications Inc.

Longman (1978) *Longman dictionary of contemporary english*. London: Longman Group Limited.

Loosemore, M. (1998a). The three ironies of crisis management in construction projects. *International Journal of Project Management, 16*(3), 139–144.

Loosemore, M. (1998b). The influence of communication structure upon crisis management efficiency. *Construction Management and Economics, 16*, 661–671.

Loosemore, M. (1998c). Reactive crisis management in construction projects—patterns of communication and behavior. *Journal of Contingencies and Crisis Management, 6*(1), 23–34.

Loosemore, M. (2000). *Crisis Management in Construction Projects*. Reston: ASCE Press.

Lorenz, E. N. (1963). Deterministic nonperiodic flow. *Journal of Atmospheric Sciences, 20*, 130–141.

Lorenz, E. N. (1993). *The essence of chaos*. Seattle: University of Washington Press.

Low, S. P., David, K. H. Ho, & Yeap, S. A. (1999). Crisis management: A survey of property development firms. *Property Management, 17*(3), 231–251.

Low, S. P., Liu, J. Y., & Stephen, S. (2008). Attitudes towards business continuity management. *The Singapore Engineer*, 36–44.

Low, S. P., & Jiang, H. B. (2003). Internationalization of Chinese construction enterprises. *Journal of Construction Engineering and Management, 129*(6), 589–598.

Low, S. P., & Shi, Y. (2001). Cultural influences on organizational processes in international project: Two case studies. *Work Study, 50*, 267–285.

Lu, C., & So, R. W. (2005). Return relationships between listed banks and real estate firms: Evidence from seven Asian economies. *The Journal of Real Estate Finance and Economics, 31*(2), 189–206.

Luhmann, N. (1995). *Social systems*. Stanford: Stanford University Press.

Luhman, J. T., & Boje, D. M. (2001). What is complexity science? A possible answer from narrative research. *Emergence, 3*(1), 158–168.

Luthans, F., Hodgetts, R. M., & Rosenkrantz, S. A. (1988). *Real managers*. Cambridge: Ballinger.

Mackenzie, H. (1966). Structural centrality in communication networks. *Psychometrika, 31*, 17–25.

Markovsky, B. (1998). Social networks and complexity theory. Presentation transcript. In *International Conference on Social Networks and Social Capital*, Duke University, Nov. 1, 1998.

Magnus R (2005), *Final Report of the Public Inquiry into the Nicoll Highway Collapse*, 20 April 2004, chairman Judge Richard Magnus, Ministry of Manpower, Singapore.

Marlow, E., & Wilson, P. (1997). *The breakdown of hierarchy: Communicating in the evolving workplace*. Boston: Butterworth-Heinemann.

Maturana, H. (1988). Everything is said by an observer. In W. Thompson (Ed.), *Gaia, a way of knowing: Political implications of the new biology* (pp. 65–82). San Francisco: Lindisfarnes Press.

McKelvey, B. (2003). Emergent order in firms: Complexity science vs. the entanglement trap. In E. Mitleton-Kelly (Ed.), *Complex systems and evolutionary perspectives on organizations: The application of complexity theory to organizations* (pp. 99–125). New York: Pergamon Press.

McMillan, E. (2006). *Complexity, organizations and change—an essential introduction*. USA: Routledge.

Millar, D. P., & Heath, R. L. (2004). *Responding to crisis, a rhetorical approach to crisis communication*. London: Lawrence Erlbaum Associates.

Minai, A. T. (1984). *Architecture as environmental communication*. Berlin: Mouton.

Mindszenthy, B. J., Watson, T. A. G., & Koch, W. J. (1988). *No surprise: The crisis communications management system*. Toronto: Bedford House Publishing.

Mitleton-Kelly, E. (2005). Ten principles of complexity and enabling infrastructures. In E. Mitleton-Kelly, (Ed.), *Complex systems and evolutionary perspectives on organisations: The application of complexity theory to organisations* (Chap. 2, pp. 23–50). Pergamon.

Mitroff, I. I. (1988). Crisis management: Cutting through the confusion. *Sloan Management Review, 29*(2), 15–21.

Mitroff, I. I., & Pearson, C. M. (1993). *Crisis management: A diagnostic guide for improving your organization's crisis-preparedness*. San Francisco: Jossey-Bass.

Mitroff, I. I. (1994). Crisis management and environmentalism: A nature fit. *California Management Review, 36*(2), 101–113.

Mitroff, I. I., & Anagnos, G. (2001). *Managing crises before they happen*. NewYork: AMACOM.

Mitroff, I. I. (2005). *Why some companies emerge stronger and better from a crisis: 7 essential, lessons for surviving disaster*. New York: AMACOM.

Mintzberg, H. (1980). *The nature of managerial work*. Engewood Cliffs: Prentice Hall.

Moffat, J. (2003). Compleixty theory and network centric warfare. CCRP Publication Series. Retrieved August 23, 2007, from http://www.dodccrp.org/files/Moffat_Complexity.pdf.

Moore, R. M., & Dainty, A. R. J. (2001). Intra-team boundaries as inhibitors of performance improvement in UK design and build projects: A call for change. *Construction Management and Economics, 19*, 559–562.

More, E. (1995). Crisis management and communication in Australian organizations. *Australian Journal of Communication, 22*, 31–47.

Morgan, G. (1997). *Images of organization* (2nd ed.). Thousand Oaks: Sage.

Murphy, P. (2000). Symmetry, contingency, complexity: Accommodating uncertainty in public relations theory. *Public Relations Review, 26*(4), 447–62.

Newcombe, R. (1996). Empowering the construction project team. *International Journal of Project Management, 14*(2), 75–80.

Newcombe, R. (1997). Procurement paths—a cultural/political perspective. In: C. H. Davidson, & T. A. Meguid, (Eds), *Procurement—a Key to Innovation. Proceeding of CIB W92*, Montreal: IF Research Corporation.

Ocal, E., Oral, E. L., & Erdis, E. (2006). Crisis management in Turkish construction industry. *Building and Environment, 41*, 1498–1503.

Ofori, G. (1994). Formulating a long-term strategy for developing the construction industry of Singapore. *Construction Management and Economics, 12*, 219–231.

Ofori, G. (2003). Frameworks for analyzing international construction. *Construction Management and Economics, 21*, 379–391.

Ofori, G., & Debrah, Y. A. (1998). Flexible management of workers: Review of employment practices in the construction industry in Singapore. *Construction Management and Economics, 16*(4), 397–408.

Ormerod, P. (1999). *Butterfly economics: A new general theory of economic and social behaviour*. London: Faber and Faber.

Packard, N. H. (1988). Adaptation toward the edge of chaos. In J. A. S. Kelso, A. J. Mandell, & M. F. Shlesinger (Eds.), *Dynamic patterns in complex system* (pp. 293–301). Teaneck: World Scientific.

Paraskevas, A. (2006). Crisis management or crisis response system? A complexity science approach to organizational crises. *Management Decision, 44*(7), 892–907.

Pascale, R. T., Millemann, M., & Gioja, L. (2000). *Surfing the edge of chaos* (1st ed.). New York: Crown Business.

Pearce, W. B. (1989). *Communication and the Human Condition*. Carbondale: Southean Illinois University Press.

Pearson, C. M., & Clair, J. A. (1998). Reframing crisis management. *Academy of Management Review, 23*(1), 59–76.

Pedler, M., Burgoyne, J., & Boydell, T. (1991). *The learning company, a strategy for sustainable development*. London: McGraw-Hill.

Perry, J. G., & Hayes, R. W. (1985). Risk and its management in construction projects. *Proceedings of Institute of Civil Engineers, 1*(78), 499–521.

Pfeffer, J. (1981). *Power in organizations*. Boston: Pitman.

PMI. (2004). *Project Management Body of Knowledge*. Pennsylvania: Project Management Institute.

Polonsky, M. J. (1995). A stakeholder theory approach to designing environment marketing strategy. *Journal of Business and Industry Marketing, 10*(3), 29–46.

Postmes, T., Tanis, M., & de Wit, B. (2001). Communication and commitment in organisations: A social identity approach. *Group Processes and Intergroup Relations, 4*(3), 207–226.

Powell, W. W. (1998). Learning from collaboration: Knowledge and networks in the biotechnology and pharmaceutical industries. *California Management Review, 40*(3), 228–240.

Stacey, R. D. (1996) *Complexity and creativity in organization*. San Francisco: BK Publisher.

Priesmeyer, H. R. (1992). *Organizations and chaos: Defining the methods of non linear management*. Westport: Quorum Books.

Prigogine, I. (1997). *The end of certainty*. New York: The Free Press.

Prigogine, I., & Stengers, I. (1984). *Order out of chaos: Man's new dialogue with nature*. New York: Bantam.

Pugh, D. S., & Hickson, D. J. (1996). *Writers on organizations*. 5th edn, London: Penguin Books.

Quarantelli, E. I. (1988). Disaster crisis management: A summary of research findings. *Journal of Management Studies, 25*(4), 373–386.

Ragins, B. R. (1995). Diversity, power, and mentorship in organizations: A cultural, structural and behavioral perspective. In M. M. Chemers, S. Oskamp, M. A. Costanzo (Eds), *Diversity in organizations: New perspectives for a changing workplace* (pp. 91–132). Thousand Oaks: Sage.

Ray, S. J. (1999). *Strategic communication in crisis management lessons from the airline industry*. Westport: Quorum.

Reed, M. I. (1996). Expert power and control in late modernity: An empirical review and theoretical synthesis. *Organization Studies, 17*(4), 573–597.

Reid, J. L. (2000). *Crisis management: Planning and media relations for the design and construction industry*. New York: Wiley, Inc.

Reulle, D. (1991). *Chance and chaos*. Princeton: Princeton University Press.

Richardson, B. (1994). Socio-technical disaster: Profile and prevalence. *Disaster Prevention and Management, 3*(4), 41–69.

Richardson, K. A. (2005). *Managing organizational complex: Philosophy, theory and application*. Greenwich: Information Age Publishing.

Richardson, K. A, Fail, A., Roos, L, iij Lissack, M. R. (2005). The coherent management of complex projects and the potential role of group decision support systems. Managing organizational complexity: Philosophy, theory, and application. In K. Richardson (Ed.), *Managing the complex* (pp. 433–458). Greenwich: Information Age Publishing, Inc.

Rifkin, W., & Fulop, L. (1997). A review and case study of learning organizations. *The Learning Organization, 4*(4), 135–148.

Robbins, S. P., & Judge, T. A. (2007). *Organizational Behavior* (12th ed.). Upper Saddle River: Prentice Hall.

Robson, C. (2002). *Real world research: A resource for social scientists and practitioner—researchers* (2nd ed.). Oxford: Blackwell Publisher Ltd.

Rogers, E. M., & Agarwala-Rogers, R. (1976). *Communication in organization.* New York: Free Press.

Rogers, E. M., & Kincaid, D. (1981). *Communication networks: Toward a new paradigm for research.* New York: The Free Press.

Rosenhead, J. (2001) Complexity theory and management practice. Retrieved August 22, 2008, from http://www.human-nature.com/science-as-culture/rosenhead.html.

Schein, E. H. (1996). Culture: The missing concept in organizational studies. *Administrative Science Quarterly, 41*, 229–240.

SCJST. (2006). Qualification categorization of general contractor for Chinese construction firms. Retrieved August 13, 2009, from http://web.scjst.gov.cn/webSite/main/pageDetail.aspx?fid=DF408CBD-B90C-49A4-BE0F-536FF25DABCB&fcol=175030.

Scott, W. R. (1998). *Organization: Rational, natural and open systems* (4th ed.). Englewood Cliff: Prentice Hall.

Scudder, J. N. (1999). Influence, beliefs, appraisals, and effect: A test of appraisal theory in mediated context. *Journal of Applied Communication, 27*(3), 196–216.

Seeger, M. W. (2002). Chaos and crisis: Propositions for a general theory of crisis communication. *Public Relations Review, 28*(4), 329–337.

Seeger, M. W., Sellnow, T. L., & Ulmer, R. R. (2001). Public relations and crisis communication: Organizing and chaos. In R. L. Heath (Ed.), *Handbook of public relations* (pp. 155–165). Thousand Oaks: Sage.

Seeger, M. W., Sellnow, T. L., & Ulmer, R. R. (2003). *Communication and organizational crisis.* Westport: Praeger.

Sellnow, T. L., Seeger, M. W., & Ulmer, R. R. (2002). Chaos theory, informational needs and natural disasters. *Journal of Applied Communication Research, 30*(4), 269–292.

Senge, P. M. (1990). The leader's new work: Building learning organizations. *Sloan Management Review, 32*(1), 7–23.

Shaluf, I. M. (2007a). An overview on disasters. *Disaster Prevention and Management, 16*(5), 687–703.

Shaluf, I. M. (2007b). Disasters types. *Disaster Prevention and Management, 16*(5), 704–717.

Shannon, C. E. (1948). A mathematical theory of communication. *The Bell System Technical Journal, 27*, 379–423, 623–656.

Shannon, C. E., & Weaver, W. (1949). *The mathematical theory of communication.* Urbana: University of Illinois Press.

Shaw, M. E. (1954). Group structure and the behaviour of individuals in small groups. *Journal of Psychology, 38*(1), 139–149.

Shaw, M. E. (1957). Decision processes in communication nets. *Journal of Abnormal Social Psychology, 54*(3), 323–330.

Shaw, M. E. (1981). *Group dynamics: The psychology of small groups.* New York: McGraw-Hill.

Shrivastava, P. (1993). Crisis theory/practice: Towards a sustainable future. *Industrial and Environmental Crisis Quarterly, 7*, 23–42.

Sikich, G. W. (1993). *It cannot happen here: All hazards crisis management planning.* UK: PennWell Publishing Company.

Sillars, A., & Parry, D. (1982). Stress, cognition, and communication in interpersonal conflicts. *Communication Research, 9*(2), 201–226.
Sinclair, A., & Haines, F. (1993). Deaths in workplace and the dynamics of response. *Journal of Contingencies and Crisis Management, 1*(3), 125–138.
Skyttner, L. (1998). Some complementary concepts of communication theory. *Kybernetics: The International Journal of Systems and Cybernetics, 27*(2), 155–164.
Smircich, L., & Stubbart, C. (1985). Strategic management in an enacted world. *Academy of Management Review, 10*(4), 724–736.
Sommer, C. S., & Loch, C. H. (2004). Selectionism and learning in projects with complexity and unforeseeable uncertainty. *Management Science, 50*, 1334–1348.
Son, J., Aziz, Z., & Pena-Mora, F. (2007). Supporting disaster response and recovery through improved situation awareness. *Structural Survey, 26*(5), 411–425.
Stacey, R. (1992). *Managing chaos*. London: Kogan Page.
Stacey, R. (1996). *Complexity and creativity in organizations*. San Francisco: Berrett-Koehler.
Stacey, R. (2000). *Strategic management and organizational dynamics* (3rd ed.). London: Pearson Education.
Stacey, R. (2001). *Complex responsive process in organizations: Learning and knowledge creation*. New York: Routledge.
Stacey, R., Griffin, D., & Shaw, P. (2000). *Complexity and management: Fad or radical challenge to systems thinking?*. London: Routledge.
Stephens, K. K., Malone, P. C., & Bailey, C. M. (2005). Communicating with stakeholders during a crisis: Evaluating message strategies. *Journal of Business Communication, 42*(4), 390–419.
Stern, G. (1987). They don't know what they're doing: Employee surveys and the management of crises. *The Wyatt Communicator* New York, United States: Winter, 21–25.
Stocking, S. H. (1998). On drawing attention to ignorance. *Science Communication, 20*(1), 165–178.
Streatfield, P. J. (2001). *The paradox of control in organizations*. London: Routledge.
Stroh, U. (1998). Communication in a millennium of chaos and change. *Communicare, 17*(2), 16–41.
Suchman, L. (2000). Organizing alignment: A case of bridge-building. *Organization, 7*(2), 311–327.
Tan, W. (1995). *Research methods in real estate and construction*. Singapore: National University of Singapore.
Tan, W. (2002). *Practical research method* (2nd ed.). Singapore: Pearson Education South Asia.
Institute, Tavistock. (1966). *Independence and uncertainty—a study of the building industry*. London: Tavistock Publications.
Taylor, J. R., Cooren, F., Giroux, N., & Robichaud, D. (1996). The communicational basis of organization: Between the conversation and the text. *Communication Theory, 6*, 1–39.
Taylor, F. W. (1967). *The principles of scientific management*. New York: Norton.
Teo, M. M. M. (1998). *An investigation into the crisis preparedness of Australian construction companies. BSc paper*. Sydney: University of New South Wales.
Thayer, W. (1998). An inside look at the scary world of food safety scares. *Frozen Food Age, 46* (1), 12–38.
Thomas, W. L. (1999). *Using qualitative methods in organizational research*. Thousand Oaks: Sage Publication Inc.
Thompson, D. (1967). *Organizations in action*. New York: McGraw-Hill.
Thompson, P., & McHugh, D. (2002). *Work organizations: A critical introduction* (3rd ed.). Hampshire: Palgrave.
Thorelli, H. B. (1986). Network: Between markets and hierarchies. *Strategic Management Journal, 7*, 37–51.
Tyler, L. (2005). Towards a postmodern understanding of crisis communication. *Public Relations Review, 31*(4), 566–571.

Ulmer, R. R. (2001). Effective crisis management through established stakeholder relationships. *Management communication Quarterly, 14*(4), 590–615.

Urry, J. (2003). *Global complexity*. Cambridge: Polity Press.

Van Uden, J., Richardson, K. A., & Cilliers, P. (2001). Postmodernism revisited? Complexity science and the study of organizations. *Tamara: Journal of Critical Postmodern Organization Science, 1*(3), 53–67.

Varela, F. J. (1995). The re-enhancement of the concrete. In L. Steels & R. Brooks (Eds.), *The artificial life route to artificial intelligence: Building embodied, situated agents* (pp. 11–20). Mahwah: Erlbaum.

Waddock, S. (2001). Corporate citizenship enacted as operating practice. *International Journal of Value-Based Management, 14*(3), 237–246.

Waldrop, M. (1994). *Complexity—the emerging science at the edge of order and chaos*. New York: Simon and Schuster Publisher.

Walker, A. (2006). *Project Management in Construction* (5th ed.). Oxford: Blackwell Publishing.

Walker, D. H. T. (1997). Choosing an appropriate research design. *Construction Management and Economics, 15*, 149–159.

Walker, D. H. T., & Loosemore, M. (2003). Flexible problem solving in construction projects on the National Museum of Australia project. *Team Performance Management: An International Journal, 9*(1/2), 5–15.

Wang, Y. G., & Li, H. R. (2007). *Marketing competences and strategic flexibility in China*. New York: Palgrave Macmillan.

Weaver, P. (2007a). *Getting the 'soft stuff' right—effective communication to successful project outcomes*. Presented at PMI Global Congress, 6–9 October 2007, Atlanta, Georgia, USA.

Weaver, P. (2007b) *A simple view of 'complexity' in project management*. Presented at WPM Week, Singapore, 14–16 Nov 2007

Weick, K. E. (1983). Organizational communication: Toward a research agenda. In L. Putnam & M. Pacanowsky (Eds.), *Communication and organizations: An interpretive approach* (pp. 10–29). San Francisco: Sage.

Weick, K. E. (1993). The collapse of sense making in organizations: The Mann Gulch disaster. *Administrative Science Quarterly, 38*, 628–652.

Weick, K. E. (1995). *Sensemaking in Organizations*. Beverly Hills: Sage.

Weick, K. E. (1998). Improvisation as a mindset for organizational analysis. *Organization Science: A Journal of the Institute of Management Science, 9*, 543–556.

Wheatley, M. J. (1994). *Leadership and the new science: Learning about organization from an orderly universe* (1st ed.). San Francisco: Berrett-Koehler.

Wheatley, M. J. (2006). *Leadership and the new science: Discovering order in a chaotic world* (3rd ed.). San Francisco: Berrett-Koehler Publisher.

White, J., & Mazur, L. (1995). *Strategic communications management: Making public relations work*. Wokingham: Addison-Wesley.

Whyte, W. F. (1984). *Learning from the field: A guide from experience*. Beverly Hills: Sage Publication, Inc.

Wild, A. (2002). The unmanageability of construction and the theoretical psycho-social dynamics of projects. *Engineering, Construction and Architectural Management, 9*(4), 345–351.

Williams, D. E., & Treadway, G. (1992). Exxon and valdez accident: A failure in crisis communication. *Communication Studies, 43*, 56–64.

Williams, H. B. (1957). Some functions of communication in crisis behavior. *Human Organization, 16*, 15–19.

Williams, J. C. (1978). *Human behavior in organizations*. Cincinnati: South-Western Publishing.

Williams, T. M. (1999). The need for new paradigms for complex projects. *International Journal of Project Management, 17*(5), 269–273.

Winch, G. (1989). The construction firm and construction project: A transaction cost approach. *Construction Management and Economics, 7*(4), 331–345.

Winsor, D. A. (1990). The construction of knowledge in organizations: asking the right questions about the challenger. *Journal of Business and Technical Communication, 4*, 8–20.

Winston, T. (1997). Introduction to case study. *The Qualitative Report*, *3*(2). Retrieved December 2008, from http://www.nova.edu/ssss/QR/QR3-2/tellis1.html.

Yin, R. K. (2003). *Case study research: Design and method* (3rd ed.). Thousand Oaks: Sage Publication Inc.

Youngblood, M. D. (1997). *Life at the edge of chaos: Creating the quantum organization.* Dallas: Perceval Publishing.

Yukl, G. A. (1981). *Leadership in organizations*. Englewood Cliffs: Prentice-Hall.

Zhong, Y., & Low, S. P. (2007a). Project complexity and strategic management in construction projects. *Proceedings of the 3rd ASEAN PGS 2007*, 200–207, University of Malaya, Malaysia, 2007.

Zhong, Y., & Low, S. P. (2007b). Project complexity and stakeholder management in construction projects. *Proceedings of MICRA Management in Construction Research Conference*, B-58-1-6, University of Technology Malaysia, Malaysia, 2007.

Zhong, Y., & Low, S. P. (2008). Managing stakeholder communication during a construction crisis—from the perspective of complexity. *Proceedings of CRIOCM 2008 International Research Symposium on Advancement of Construction Management and Real Estate* (pp. 319–324). Beijing.

Zhong, Y., & Low, S. P. (2009). Managing crisis response communication in construction projects—from a complexity perspective. *Journal of Disaster Prevention and Management, 18*(3), 270–282.

Zhong, Y., & Low, S. P. (2011). A complexity approach for crisis response communication management. *Proceedings of the International Construction Business and Management Symposium*, Malaysia, 2011

Zsambok, C. E. (1997). Naturalistic decision making research and improving team decision making. In C. E. Zsambok & G. Klein (Eds.), *Naturalistic decision making* (pp. 111–120). Mahwah: Erlbaum.

About the Authors

Zhong Ying is presently working for the Shanghai Urban and Rural Construction and Transportation Development Research Institute, China with senior management responsibilities for urban construction, policy research and formulation, as well as consultation. Before this present appointment, he has worked for Chinese enterprises in real estate development and construction as well as a project management consultant in various parts of China. He holds a M.Sc. (Eng) degree from the University of Birmingham and a Ph.D. from the National University of Singapore.

Low Sui Pheng is presently a Professor of Building in the School of Design and Environment, National University of Singapore where he has previously served as Head and Vice-Dean. He teaches construction project management and has published, researched, and consulted extensively both in Singapore and overseas. A Fellow of the Chartered Institute of Building, he holds a D.Sc. from the University of Birmingham and a Ph.D. from University College London.

Z. Ying and L. Sui Pheng, *Project Communication Management in Complex Environments*,
DOI: 10.1007/978-981-4560-64-1, © Springer Science+Business Media Singapore 2014

MIX
Papier aus verantwortungsvollen Quellen
Paper from responsible sources
FSC® C105338

If you have any concerns about our products,
you can contact us on
ProductSafety@springernature.com

In case Publisher is established outside the EU,
the EU authorized representative is:
Springer Nature Customer Service Center GmbH
Europaplatz 3, 69115 Heidelberg, Germany

Printed by Libri Plureos GmbH
in Hamburg, Germany